The Challenge of Deepwater Terminals

The Challenge of Deepwater Terminals

Louis K. Bragaw
United States Coast Guard Academy

Henry S. Marcus
Massachusetts Institute of Technology

Gary C. Raffaele
University of Texas at San Antonio

James R. Townley
United States Coast Guard Headquarters

Lexington Books
D.C. Heath and Company
Lexington, Massachusetts
Toronto   London

Library of Congress Cataloging in Publication Data

Main entry under title:

The Challenge of deepwater terminals.

   Bibliography: p.
   1. Petroleum shipping terminals—United States. I. Bragaw, Louis K.
HE553.C45   387.1'53   73-11661
ISBN 0-669-88526-6

To those who make it
all worthwhile:
Kathy, Edwina, Virginia, Sueanne

Contents

List of Figures

List of Tables

Preface

A number of new developments in the maritime industry . . . now require from us new and imaginative and innovative responses. These new developments require nothing less than change. . . . We must get ready to deal with perhaps the most spectacular change of all—the advent of the very large crude carriers (VLCC) and the offshore ports. The sheer awesome size of these vessels plus their potential for danger demand that we re-examine every phase of their operation, their maintenance, and the skills employed aboard.[a]

These remarks of Secretary of Transportation Coleman underscore *The Challenge of Deepwater Terminals*, and serve as rationale for this book. After researching this area, the authors came to the opinion each from his own perspective, that a need existed to distill the voluminous yet scattered array of information needed to study deepwater terminals. It is the objective of the authors to put in one book of readable length the key factors derived from published data, and the salient issues on the subject derived from the authors' own research. This has been attempted from the economic, social, political, technological and environmental viewpoints in order to develop some policy and management ideas on the subject. In addition, the authors have tried to extensively reference this work for the benefit of readers who may wish to do further study in the area.

---

[a]These remarks were delivered by the Honorable William T. Coleman, Jr., Secretary, United States Department of Transportation, as part of his Keynote Address at the *National Symposium on Marine Transportation Management*, Philadelphia, Pa., April 29, 1975.

Acknowledgments

For many years the authors have benefited from the opportunity to discuss the various issues involved with deepwater terminals with a myriad of people in industry, government, labor and academia. It would be impossible to list the many individuals; however, the authors sincerely appreciate the generous help given them in many ways.

In particular, the authors would like to thank Edythbelle Vail of the United States Coast Guard Academy, who provided fine secretarial assistance in preparing the manuscript. Thanks also go to the cadets in the Coast Guard Academy course in *Transportation Policy and Planning* who critiqued rough drafts of the manuscript. Appreciation is also due Olivia Rodriguez of the University of Texas at San Antonio for her work in producing the artwork appearing in the figures in this book. Thanks also go to Shirley End, Production Manager of Lexington Books, D.C. Heath and Company for work in the production phase of the book.

Although this work has received the support of so many from industry, government, labor and academia, and from the individuals specifically mentioned, all of whom deserve much credit for any positive aspects the book may present, the ideas and opinions presented are the work and personal opinions of the authors and not the organizations they represent.

East Lyme, Connecticut

May 17, 1975

Glossary of Terms

AFL	American Federation of Labor
AMA	American Maritime Association
AMO	Associated Maritime Officers
ARA	American Radio Association
Barrel	A unit of petroleum quantity, equaling 42 gallons
BMO	Brotherhood of Marine Officers
BTU	A unit of energy, by definition a British Thermal Unit
CALM	Catenary Anchor Lag Mooring, a type of Single-Buoy Mooring
CEQ	Council on Environmental Quality
CIO	Congress of Industrial Organizations
DOC	Department of Commerce
DOD	Department of Defense
DOI	Department of Interior
DOT	Department of Transportation
DWT	Deadweight tonnage—the total weight in long tons (2240 lbs.) that a tanker carries on a specified draft including cargo, fuel, water in tanks, stores, baggage, passengers, crew, and all their effects, but excluding water in the boilers
ILA	International Longshoremen's Association
IMCO	Intergovernmental Maritime Consultative Organization
ITIA	International Tankers Indemnity Association
Intertanko	International Association of Independent Tanker Owners
Jones Act	Merchant Marine Act of 1920
MARAD	Maritime Administration
MBD	A rate of energy supply, equaling one million barrels a day
MEBA	Marine Engineers Beneficial Association
MITAGS	Maritime Institute of Technology and Graduate Studies
MMP	International Organization of Masters, Mates, and Pilots
NEPA	National Environmental Policy Act of 1970
NMU	National Maritime Union
NPC	National Petroleum Council
OCAS	Oil, Chemical and Atomic Workers
OCS	Outer Continental Shelf
PAD	Petroleum Administration for Defense District
Quad	A unit of energy, equaling one quadrillion BTU
Reserves	Estimates of supply based on detailed geologic evidence
RFR	Required Freight Rate
ROU	Radio Officers' Union
SALM	Single-Anchor Leg Mooring
SIU	Seafarers International Union

SPM	Single-point mooring
Spot Rate	Single voyage charter rate
TAPS	Trans-Alaska Pipeline
Time Rate	Extended voyages, often twenty years, charter rate
TOVALOP	Tanker Owner's Voluntary Agreement Concerning Liability of Oil Pollution
TSC	Tanker Service Committee
T-2	A designation for a tanker design, usually 16,600 tons, 524 feet length, 68 feet beam, 30 feet draft
ULCC	Ultra large crude carriers (over 400,000 tons)
VLCC	Very large crude carriers (over 200,000 tons)
Worldscale	A standard tanker rate schedule, for chartering

The Challenge of Deepwater Terminals

1

The Demand for Petroleum Imports to the United States and the Challenge of Deepwater Terminals

The Energy Dilemma, Supertankers, and Deepwater Terminals

As the world enters the last quarter of the twentieth century sensitivities have increased to the dependences on energy that have developed as national economies have matured and become intertwined. The ongoing conflict in the Middle East, as spotlighted by the 1973 war, has dramatically focused attention on existing imbalances between energy-rich countries and countries dependent on energy imports to fuel their economies.

This national imbalance between energy production and consumption is part of the growing dependence on petroleum as a fuel. This dependence grew in the years since World War II as the oil fields of the Middle East developed. It continues as new reserves are discovered at increasing distances from areas of high petroleum consumption. These growing distances are particularly significant in the four-stage petroleum industry. The four-stage process of crude production, crude transportation, product refining, and product transportation is very sensitive to transportation costs. This is so because transportation connects the production and refinery stages, and transportation again connects refinery and final product distribution.

Tankers have always been a significant and economical form of petroleum transportation; tanker use accelerated after World War II because of the growing distances between production and consumption and because of increased aggregate demand. Scale economies in tanker construction and operation led to larger and larger tankers;[a] by the mid-1960s supertankers of 150,000 deadweight (DWT) tons were being constructed.[b] As the petroleum industry tended to center the refinery stage near either oil field production or final distribution centers to minimize transportation costs, the existence of supertankers presented a definite challenge to nations involved in petroleum trade to provide deepwater ports and gain entry to the larger tanker transportation system.

Specifically, the deepwater terminal challenge is one of system entry.

[a]Some feel that the closing of the Suez Canal in 1967 was a significant factor in encouraging the growth of supertankers; others feel that economies of scale with supertankers made their use inevitable.

[b]The term "supertanker" is used in this book to describe a tanker of approximately 150,000 DWT or over. The term "very large crude carrier" (VLCC), refers to tankers of about 200,000 DWT and over, while "ultra large crude carrier" (ULCC) denotes tankers of about 400,000 DWT and over.

1

Supertanker drafts began to bar them from entry to many conventional ports of limited water depths. Deepwater terminals, described in detail in Chapter 3, spring up of necessity to service the larger tankers. The most common system eventually became the single point mooring, which in simplest terms consists of a pipe extending out to a buoy moored in water deep enough to allow a tanker to pump cargo. But insufficient water depths are not only physical barriers to entry; to the extent that they bar tankers from delivery they are also economic barriers to entry. In the United States some ports had deepened channels and harbors and altered facilities, but at the end of 1974 no legislation had been passed setting national priorities for deepwater terminals.

Scale economies possible in filling and discharging oil brought about a *de facto* international system for marine petroleum transportation consisting of supertankers and deepwater terminals. If the United States is to become a part of this system by establishing deepwater terminals, the terminals should be assessed from the economic, environmental, and labor perspectives; their onshore impact should certainly be considered. In addition to traditional corporate concerns for minimizing transportation distances, and hence costs, society has increasingly questioned whether energy in its many resource forms—petroleum, coal, nuclear fission, or other—may be transformed to desirable energy without at the same time producing deleterious side effects, or "externalities." Tankers, with half the petroleum transportation market, move not only "dirty" cargo, the industry term for crude oil, but also "clean" product, as gasoline and other refined products are called. The *Torrey Canyon* grounding, and several other incidents, dramatized that there is nothing very aesthetic or recycling in nature about oil spills and the resulting pollution of the marine environment.

As environmental concerns grow, each potential energy source is undergoing extensive worldwide examination. Environmental and safety questions have surrounded attempts to increase coal and nuclear energy development, and regulatory questions have been encountered with natural gas. In the United States nearly 45 percent of the energy consumed has come from petroleum. Future additional demands could be met in part by supply from Outer Continental Shelf operations,[c] from Trans-Alaska Pipeline construction,[d] or from increased petroleum imports. All of these supply options face many questions; this book will focus on the challenge the import option presents in the United States to marine petroleum transportation, i.e., to tankers and deepwater terminals. Before further exploring the challenge one should first review the relative size of world and domestic energy systems, in order to place the energy flows that will be considered in perspective.

---

[c]A fine assessment of Outer Continental Shelf oil and gas operations and their potential as a supply option is provided in Don E. Kash, Irvin L. White, et al.: *Energy Under the Oceans* (Norman, Okla.: The University of Oklahoma Press, 1973).

[d]For the definitive study, see the Impact Statement, United States Department of the Interior; "An Analysis of the Economic and Security Aspects of the Trans-Alaska Pipeline" (Washington, D.C., December 1971).

A General Energy Overview

Energy, or work in scientific measurement, is usually measured by the joule, newton-meter, calorie, ft.-lb., kilowatt hour, or British Thermal Unit (BTU). The unit commonly used in aggregate energy statistics is the quad, representing one quadrillion BTUs. Table 1-1 lists all the common energy units and conversions. To provide the reader who wishes some familiarity with calculations in these units, Table 1-2 lists the annual energy consumption in the United States in 1970 from each of the major energy sources, measured in the units usually associated with the particular source, and also in quads.

Energy consumption in the United States in the late 1960s and early 1970s rose at rates exceeding 4 percent per annum, an increase over the average annual 3.4 percent reported for the period 1950-72.[1] A great many forecasts have been prepared for future United States energy consumption; one summary includes a range of eleven separate forecasts.[2] The rate of growth in energy demand is certainly the key factor in forecasting future energy consumption. A growth rate of 1 or 2 percent will bring about a modest figure for future demand; a rate of

Table 1-1
Energy Units and Common Conversions

	Energy Units		Common Conversions
1 Joule	= 1 Newton-Meter	1 Barrel	= 42 Gallons
1 Ft-Lb.	= 1.356 Joules	1 Barrel	= 0.136 Tons
1 Calorie	= 4.184 Joules	1 Ton	= 7.35 Barrels
1 BTU	= $1.055 (10)^3$ Joules	1 Ton	= 308 Gallons
1 Watt-Hr.	= $3.6 (10)^3$ Joules	1 Barrel/Day	= 50 Tons/Year
1 Kw-Hr.	= 3413 BTU	1 Barrel	= $5.8 (10)^6$ BTU
$1 (10)^9$ BTU	= 1 Billion BTU	1 Cu.Ft. Gas	= 1031 BTU
$1 (10)^{12}$ BTU	= 1 Trillion BTU	1 Ton Lignite Coal	= 20 to $40 (10)^6$ BTU
$1 (10)^{15}$ BTU	= 1 Quad (1 Quadrillion BTU)	1 Ton Uranium 233	= $5.8 (10)^{13}$ BTU

Table 1-2
Energy Consumption by Source in the United States in 1970—Sample Calculations

Energy Source	Energy Consumption		
Petroleum	$14.7(10)^6$ Barrels/Day	= $30(10)^{15}$ BTU/Yr	= 30 Quads
Natural gas	$59.5(10)^9$ Cu.Ft./Day	= $21(10)^{15}$ BTU/Yr	= 21 Quads
Coal	$525(10)^6$ Tons/Year	= $16(10)^{15}$ BTU/Yr	= 16 Quads
Hydroelectric, etc.			= 1 Quad
Yearly consumption			68 Quads

4.2 percent, the "base case" in one study, will obviously dictate a much larger future energy demand.[3]

With a domestic growth rate of 3.4 percent, or greater, per annum, United States aggregate domestic consumption has still declined as a fraction of world consumption from 44 percent in 1950 to 33 percent by 1968. Even this is a huge per capita share when one considers that United States population constitutes 6 percent of world population. The relative decline in United States domestic consumption has occurred because of spreading geographic use, and rises in absolute demand worldwide. Table 1-3 lists the quads used by each consuming section in the United States in 1968, along with the growth rates in *each*. As 61 quads were consumed in the United States, and domestic consumption represents a third of world consumption, approximately 183 quads were consumed worldwide in 1968. Preliminary reports of the Bureau of Mines indicate that United States energy consumption had risen to 75 quads for 1973.

Although the world consumption rate is increasing, it can be placed in perspective by comparing it with Table 1-4, which provides a summary of major energy resources in the United States. Table 1-5 summarizes major worldwide energy resources. The term "reserves" as used here will refer to estimates based on "detailed geologic evidence, usually obtained through drilling," while "recoverable resources" and "remaining resource base" reflect estimates with "less detailed knowledge and more geologic inference." In Tables 1-4 and 1-5, *resources* refer to the amount of the fuel in the ground, including that which has not yet been discovered; *reserves* are those resources that have been delineated

Table 1-3
Total and Sectoral Energy Consumption in the United States, 1968

Consuming Sector	Consumption (Quads)	Percent of Total	Growth Rate
Residential	11.6	19.2	4.8%
Commercial	8.8	14.4	5.4%
Industrial	25.0	41.2	3.9%
Transportation	15.2	25.2	4.1%
Total	60.5[a]	100.0	4.3%[b]

[a]The 1968 consumption of 60.5 quads does not add due to rounding. This figure compares with U.S. consumption figures of 61.7 quads in 1968, 67.1 quads in 1970, and 72.1 quads in 1972, appearing in *Exploring Energy Choices*, A Preliminary Report of the Energy Policy Project of the Ford Foundation, 1974, p. 74.

[b]The consumption growth rate may have been rising at 4.3 percent, but in 1975 is probably rising at a rate closer to the growth rates reported from 1955 to 1970. The *Guide to NPC Report on United States Energy Outlook*, 1972, p. 6, reported a growth rate for 1955 to 1970 of 3.6 percent per annum. The preliminary report of the Energy Policy Project of the Ford Foundation, *Exploring Energy Choices*, 1974, p. 41, discusses an energy consumption growth in the United States of 3.4 percent per year for the period 1950-1972.

Source: *Patterns of Energy Consumption in the United States*, Report for the Office of Science and Technology by Stanford Research Institute, January 1972, p. 5.

Table 1-4
Summary of Major Energy Resources, United States, in Quads

	1973 Consumption (Quadrillion BTU)	Cumulative Production (Q BTU)	Reserves (Q BTU)	Recoverable Resources (Q BTU)	Remaining Resource Base (Q BTU)
Petroleum	34.7	605	302	2,910	16,790
Shale oil	—	—	(465)	N/A	975,000
Tar sands	—	—	—	N/A	168
Natural gas	23.6	405	300	2,470	6,800
Coal	13.5	810	4,110	14,600	64,000
Strippable coal	N/A	N/A	925	2,600	2,600
Low-sulfur coal	N/A	N/A	2,390	N/A	38,200
Uranium					
Used in light-water reactors	.85	2	228	600	3,200
Used in breeders	—	—	17,700	47,000	200,000,000
Thorium used in breeders	—	—	4,200	17,500	570,000
Hydropower	2.9			5.8*	

N/A not available
* ultimate capability

Source: *Exploring Energy Choices*, Preliminary Report of the Energy Policy Project of the Ford Foundation, 1974, as reported in table 3, p. 74. Reprinted with permission of the Energy Policy Project of the Ford Foundation.

Table 1-5
Summary of Major Energy Resources, World, in Quads

	Cumulative Production (Q BTU)	Reserves (Q BTU)	Recoverable Resources (QBTU)	Remaining Resource Base (QBTU)
Petroleum	1,550	3,680	14,400	60,000
Shale oil	–	1,100	N/A	12,000,000
Tar sands	.6	1,000	2,150	N/A
Natural gas	670	1,860	15,800	32,000
Coal	3,340	N/A	N/A	340,000
Uranium				
Used in light-water reactors	N/A	510	990	650,000,000
Used in breeders	–	40,000	77,000	600 billion
Thorium used in breeders	–	22,000	66,000	N/A

Source: *Exploring Energy Choices*, Preliminary Report of the Energy Policy Project of the Ford Foundation, 1974, as reported in table 2, p. 74. Reprinted with permission of the Energy Policy Project of the Ford Foundation.

and are capable of being developed for production; and *supplies* are the quantities that could be produced per day or per year.[4] There are some differences in these concepts between various energy sources, and certainly between studies. In the words of the National Petroleum Council, "it is still possible to make relevant comparisons regarding the resource base and supply capabilities of individual fuels."[5]

Reserves are, of course, hard to measure; world pressures, price elasticities, and changing technologies can well alter the estimates. There have been many interesting discussions of United States and world petroleum reserves. The environmental impact statement, *Analysis of the Economic and Security Aspects of the Trans-Alaska Pipeline*, provides an excellent example.[6] Hubbert's work on "The Energy Resources of the Earth" also makes good reading in this area.[7]

In 1968 the energy consumed in the United States came from various energy sources, indicated in Table 1-6. They were consumed in that year by the consuming sectors listed in Table 1-3. Table 1-6 underscores the dominant role of petroleum as an energy source in the United States; oil supplied nearly 45 percent of domestic consumption in 1968. Yet by examining Table 1-7 one sees that petroleum reserves are finite in size when compared to production rates. World petroleum reserves, reported at various levels, and in Table 1-7 equaling approximately 3600 quads, are skewed considerably geographically. Over 56 percent of reported reserves are in the Middle East; 21 percent are in Saudi Arabia, while another 10 percent are in Iran and Kuwait. This contrasts with

Table 1-6
Distribution of United States Energy Consumption, by Energy Source, 1968

Fuel Source Consumed	Percent of Total
Petroleum	44.2
Natural gas	32.3
Coal	22.0
Hydro	1.3
Nuclear	0.2

Source: *Patterns of Energy Consumption in the United States*, Report for the Office of Science and Technology by Stanford Research Institute, January 1972, p. 22.

only 5.5 percent reserves reported for the United States, including Alaska. Table 1-8 shows the geographical distribution of reported reserves.

Modeling United States Energy
Supply and Demand

Any model constructed to forecast demand for petroleum imports to the United States must consider the full range of social, political, economic, and technological factors. Both aggregate energy demand and the set of policies that might affect demand have to be considered along with the supply options that would meet demand. Demand and supply are of course not independent, and in the final analysis must equilibrate.

When United States domestic energy consumption grew at increasing rates in the late 1960s, domestic energy production did not keep pace. While domestic production had been reported as growing at 3 percent from 1950 to 1970, growth "has been at a virtual standstill since then."[8] In the late 1960s and early 1970s imported petroleum has to a large extent filled the gap between domestic consumption and production, and has been a driving force in the expansion in tanker use. To measure the scope of the marine petroleum transportation system that will be necessary to transport imported petroleum to the United States, one should examine some of the aggregate models that have been constructed.

In 1970, the National Petroleum Council, an industry advisory board to the Secretary of the Interior, was asked to undertake a comprehensive examination of the energy outlook for the United States. In July 1971, the National Petroleum Council issued an interim report of its Committee on United States Energy Outlook, chaired by John G. McLean, Chairman and Chief Executive Officer of Continental Oil Company. The final report, *U.S. Energy Outlook,*[e]

[e]For the complete study, see Committee on U.S. Energy Outlook of the National Petroleum Council, "U.S. Energy Outlook" (Washington, D.C., 1972). This study will be discussed in the following paragraphs.

Table 1-7

World Petroleum Production and Reserves (in million barrels per day)

Area	1971	1972	1973	Sept. 1973	Reserves Jan. 1, 1974 (in billions of barrels)
Middle East					
Abu Dhabi	0.9	1.0	1.3	1.4	21.5
Iran	4.5	4.9	6.0	5.9	60.0
Iraq	1.7	1.5	1.9	2.0	31.5
Kuwait	2.9	2.8	2.9	3.2	64.0
Saudi Arabia	4.5	5.7	7.4	8.3	132.0
Other		1.2	1.9		41.2
Total Middle East		17.6	21.4		350.2
Africa					
Algeria	0.8	1.1	1.0	1.1	7.6
Libya	2.8	2.2	2.1	2.3	25.5
Nigeria	1.5	1.8	2.0	2.0	20.0
Other		0.6	0.6		14.2
Total Africa		5.6	5.8		67.3
Asia-Pacific					
Indonesia	0.9	1.0	1.3		10.5
Other			0.9		5.1
Total Asia-Pacific			2.2		15.6
Europe			.4		16.0
Western Hemisphere					
U.S.	9.5	9.5	9.2		34.7
Canada	1.4	1.5	1.8		9.4
Venezuela	3.6	3.2	3.4		14.0
Other		1.3	1.8		17.7
Total Western Hemisphere		15.7	16.1		75.8
Communist nations			9.3		103.0
Total world			55.2		627.9[a]

[a]World Petroleum reserves of 627.9 billion barrels has an energy equivalent of 3,640 quads; in agreement reserves in quads listed for the world in table 1-5.

Source: *Exploring Energy Choices*, Preliminary Report of the Energy Policy Project of the Ford Foundation, 1974, as reported in table 6, p. 76. Reprinted with permission of the Energy Policy Project of the Ford Foundation.

Table 1-8
The World of Known Oil Reserves, 1973

Area	Percent	Area	Percent
Middle East	56%	Saudi Arabia	21%
		Kuwait	10%
		Iran	10%
		Iraq	5%
		Other	10%
Communist Countries	16%		16%
Western Hemisphere	12%	United States	5.5%
		Canada	1.5%
		Other	5.0%
Europe	2.5%		2.5%
Africa	11%		11%
Indonesia	1.5%		1.5%
Other	1%		1%
Total	100%		100%

Source: *Exploring Energy Choices*, A Preliminary Report of the Ford Foundation's Energy Policy Project, 1974, p. 17. Reprinted with permission of the Energy Policy Project of the Ford Foundation.

was published in December 1972.[9] In December 1971, the trustees of the Ford Foundation authorized an Energy Policy Project "to explore the whole complex of energy issues facing the nation."[10] The project, headed by S. David Freeman, and consisting of a staff of economists, engineers, scientists, writers, lawyers, and outside consultants, published their interim report, entitled *Exploring Energy Choices*,[f] in the spring of 1974. In the introduction to this report, Freeman stated that the project is "essentially a study of U.S. energy policy, but [recognizes] that U.S. policy must reflect conditions in the rest of the world."[11]

Both the National Petroleum Council study and the Energy Policy Project of the Ford Foundation are large studies that have considered models capable of demand forecasts of petroleum imports to the United States. The key to these models are the demand forecast for energy, and the supply mix hypothesized to meet demand. The growth, or lack of it, in an energy demand forecast has become a topic of a great deal of discussion. (The rate of energy growth hypothesized determines the energy demand projected for any given year. The

[f]For the preliminary report, see the Energy Policy Project of the Ford Foundation, "Exploring Energy Choices: A Preliminary Report" (Washington, D.C., 1974). This study will be discussed in the following paragraphs; along with the National Petroleum Council study, the reports serve as a basis for the statistics in this chapter.

low growth rate of the Energy Policy Project "zero energy growth" future was criticized as unrealistic by William P. Tavoulareas, President of Mobil Oil Corporation and a member of the Advisory Board of the Project.) A high growth rate assumption will probably cause a model forecast of high imports; a low growth rate in energy demand will mean import levels will be modest. Demand can be affected by a myriad of factors; including gross products, government policies, energy-consuming technologies, environmental controls, price, and availability.[g] Each particular combination of these factors, and the factors affecting supply, will describe a particular situation that may be encountered.

Both the National Petroleum Council study and the Energy Policy Project defined several "scenarios," or alternate futures. Each scenario is selected from an infinite number of scenarios that could be selected to portray the future. A particular one, based on its set of assumptions, provides a tool for policy analysis of the various factors that would bring about the situation. It also provides a framework for considering the many ramifications that the occurrence of the scenario would bring into being. For example, the National Petroleum Council Study assumed three growth rates; 4.4 percent, 4.2 percent, and 3.4 percent, for their three projections of aggregate energy demand to 1985.[12] The Ford Energy Policy Project in their interim report assumed three different growth rates. They varied each of these over the time frames 1975-85 and 1985-2000. The growth rates used in the 1975-85 interval were 3.5, 1.7, and 1.4 percent.[13]

Alternate supply cases comprise the other half of energy supply-demand scenarios. The NPC study defined four alternate futures for energy supply, which they termed Cases I, II, III, and IV. Case I, presents the most optimistic set of supply options, supposing great effort to develop domestic fuel sources. It assumes that additions to reserves (the result of both finding rate and drilling rate) increase to an average of 3.5 billion barrels annually; that all new electrical generating plants will be nuclear; that coal production will be increased by 5 percent; and that other fuels will be developed and produced without any environmental, political, or economical problems. Case II is a less optimistic scenario. It supposes slightly lower energy production in each of the categories. Case III reflects some improvement over energy supply in the early 1970s. It assumes that "reserves will be slightly increased, that nuclear expansion will occur consistent with the AEC's most favorable forecasts, and that coal and new sources will be developed at the same rates as Case II." Case IV, the lowest supply scenario, assumes that coal production will remain essentially constant and that present dilemmas in siting and licensing of nuclear power will continue. It also supposes that environmental constraints will retard the development of resources.[14]

The Energy Policy Project developed three scenarios, reflecting a wide range

gFor a description of the four factors deemed to be most significant long range determinants of energy demand see the "Guide to the National Petroleum Council Report on United States Energy Outlook," The National Petroleum Council, (Washington, D.C., 1972), p. 6.

of alternate futures. Their first scenario, termed "historic growth," assumes that the demand for energy will grow at a rate of approximately 3.5 percent, and that the government "will not deliberately impose any policies that might effect . . . ingrained habits of energy use." On the supply side the historical growth scenario assumes that the government will adopt a firm posture favoring development of all possible supplies. Addressing the question of oil imports, the study stated: "If it proved feasible to increase oil imports on a large scale, then the pressure on domestic resources would relax somewhat." This scenario recognizes the dilemmas between vigorous development of domestic resources and the alternative of filling demand gaps by imported petroleum.

Their second scenario, labeled "technical fix," reflects on the demand side "a determined, conscious national effort to reduce demand for energy through the application of energy-saving technology." It allows energy demand to rise at approximately 1.7 percent in the 1975-1985 interval, and at still lower rates in 1985-2000. Energy growth at this level still obviously allows a great deal more flexibility in the supply options, given the mix of energy sources already discussed. Imported petroleum would be necessary under this scenario in 1985, in varying amounts depending on the rate of nuclear-power development. However, no petroleum imports would be necessary by the year 2000 at the lower energy growth rate, even in the low nuclear development case.

The third future, "zero energy growth," provides a marked contrast to the previous two scenarios. It represents the alternate future in which society becomes so concerned about social and environmental costs that new patterns of energy growth would come into being. Zero energy growth would represent an area where "durability, not disposability of goods" would be emphasized; where the notion "more is better" would be replaced by an ethic where "enough is best."[15] This scene could be further envisioned through Boulding's descriptions of an emerging "spaceship economy," replacing the "cowboy economy."[16] Under these conditions the demand for petroleum imports would be virtually nonexistent. Such conditions, or the threat of them, could have an effect on international crude prices.

The Energy Policy Project stated that the scenarios all use standard demographic forecasts; they modify some forecasts as may be necessary in the technical fix and zero energy growth scenarios. As in each of the four National Petroleum Council cases, the three Energy Policy Project scenarios have to assume certain overall conditions, hypothesize a demand level, assess a myriad of potential supply sources, and hypothesize a mix of technologies that will meet the particular supply level required. For example, in the historical growth scenario of the Energy Policy Project, 1972 is used as a base year, and supply options are developed for "high import," "high fossil fuel," and "high nuclear" supply mixes for the years 1980, 1985, and 2000. The supply mix for the technical fix future utilizes 1980 as a base year, and goes on to provide a "base," "high nuclear" and a "low nuclear" mix for 1985 and 2000.[17]

In both the Energy Policy Project and the National Petroleum Council Study the gestation periods for the various supply options were considered in achieving the supply-demand energy balances in each scenario. Considerable time lags are natural in a heavy industrial system that includes not only refineries and pipelines but also the technologies for producing newer and more complex energy supplies. The National Environmental Policy Act of 1969, and further national, state, and local legislation, have all tended to lengthen these lags and build considerable delay into any changes and adjustments in the supply system.

Much of the discussion of these scenarios centers around the choice of a growth rate in energy demand, and around the practicality of the supply mix hypothesized to meet the demand. The president of Mobil Oil Corporation, William P. Tavoulareas, was the only oil company executive on the Advisory Board to the Energy Policy Project. He concluded his dissent to the final report of the Energy Policy Project as follows:

We should ask ourselves which course carries the greater risk. If the assumptions behind the low growth cases are wrong, the result will be energy scarcity, high energy prices, unemployment, and other economic and social dislocations. On the other hand, if the assumptions supporting the case for increased supplies are wrong, we will have energy surplus and low prices.[18]

Each of the cases of the National Petroleum Council or the Energy Policy Project required a different level of petroleum imports; the growth rate of energy demand and the supply mix hypothesized are critical in forecasting a level of imports.

To examine the potential role of deepwater terminals for the United States the models of energy supply and demand discussed are invaluable in forecasting the potential demand for imported petroleum that would flow through the terminals. Forecasts of petroleum imports over specific time frames will help define the size and scope of the tanker and deepwater-terminal system necessary to handle petroleum throughputs.

Forecasting United States Petroleum Imports

As discussed, the preparation of a forecast for petroleum imports is extremely complex. The four alternate futures of the National Petroleum Council and the three scenarios of the Energy Policy Project forecast a range of imports. The pattern of domestic supply options available and prevailing consumption demand will very probably dictate that imported petroleum furnish the balance of supply necessary, at least in the short run.

Petroleum imports forecast under the four Cases of the 1972 National Petroleum Council Study are presented in Table 1-9. For 1980, they range from

Table 1-9
Forecast of Necessary Petroleum Imports: The Four NPC Cases[a]

| | Petroleum Imports to U.S. (million barrels per day) | | | |
Supply Case	1970	1975	1980	1985
I	3.4	7.2	5.8	3.6
II	3.4	7.4	7.5	8.7
III	3.4	8.5	10.6	13.5
IV	3.4	9.7	16.4	19.2

[a]The most recent supply/demand projections of the National Petroleum Council were published in September 1974 in a report entitled, *Emergency Preparedness for Interruption of Petroleum Imports into the United States.*
Source: *U.S. Energy Outlook,* A Summary Report of the National Petroleum Council, December 1972, p. 23.

5.8 million barrels a day (MBD) in the lowest import scenario, Case I, up to 16.4 MBD in Case IV. The high case levels are extremely unlikely because of the 1973 war, the oil embargo, and adjustments made in crude prices since that time. Table 1-10 displays a more precise breakdown of the petroleum supply and demand balance under Case II, which calls for a modest import level. This table also shows the effect of subtracting Canadian overland imports from the import forecasts of Table 1-9 to determine the net waterborne import levels for Case II over the years 1975, 1980, and 1985. Note from Table 1-10 that import level projections rise to 6 MBD by 1975, and remain very close to that level for 1980 and 1985.

Table 1-11 presents the demand projections for imports over cases I, II, and III of the 1972 National Petroleum Council Study. The range of figures of Table 1-11 are pictured in Figure 1-1. With such a wide range in import forecasts, it would be difficult to make plans for a delivery system without considerable contingencies. No sensitivity analysis which might have revealed a margin of error between the NPC and Energy Policy Project figures was undertaken, as these seven scenarios only illustrate a range of possible outcomes. The seven scenarios are not an exhaustive set; as National Energy Policy develops, the scenarios will change. Most figures available in 1975 indicate that Case II is the most likely forecast.[19]

Case II projects a total domestic petroleum demand in the United States of 23 MBD by 1985; it specifically forecasts that import levels will remain constant at approximately 6 MBD from 1975 to 1985. Figure 1-2 presents the Case II import forecasts; this forecast will be used for further analysis in Chapter 2. By examining Figure 1-2 one can see that a large marine petroleum transportation system will be necessary to supply the 6 MBD import level forecasted. Such a system will have to involve not only tankers, but a complex array of terminal facilities, refinery, storage, and transshipment facilities. Specifically, deepdraft

Table 1-10
United States Petroleum Supply/Demand Balance—Case II

Item	1970	1975	1980	1985
	(All figures in million barrels per day)			
Requirements[a]	14,716	17,551	20,513	23,068
Petroleum liquid production	11,297	10,186	12,939	13,887
Synthetic oil production	–	–	100	480
Total domestic petroleum supply	11,297	10,186	13,039	14,367
Petroleum imports	3,419	7,365	7,474	8,701
Percent of total required supply	23.2	42.0	36.4	37.7
Source of imports				
Canadian overland	766	1,275	1,925	2,750
Foreign waterborne	2,653	6,090	5,549	5,951

[a]Oil required to balance total energy demand, net of processing gain, stock change, unaccounted for crude and other hydrocarbon inputs.
Source: *U.S. Energy Outlook*, A Report to the National Petroleum Council's Committee on U.S. Energy Outlook, December 1972, p. 272.

Table 1-11
Demand for Foreign Waterborne Petroleum, NPC Study

Supply Case	Waterborne Petroleum Imports to the U.S. (million barrels per day)			
	1970	1975	1980	1985
I	2.653	5.940	3.865	0.814
II	2.653	6.090	5.549	5.951
III	2.653	7.229	8.699	10.724

Source: *U.S. Energy Outlook*, A Report to the National Petroleum Council's Committee on U.S. Energy Outlook, December 1972, pp. 272-73.

tankers will require deepwater terminals. The use of the large ocean tanker will be considered first.

The Nature of the Tanker Market
and Tanker Development

The growing demand for crude and refined product at increasing distances from the source of supply has led to a growth in the number of tankers in use, and an acceleration in their absolute size. Originally, petroleum imports to the United States came from Latin America, and the trips were shorter than the long hauls required from the Persian Gulf to the East and Gulf coasts.

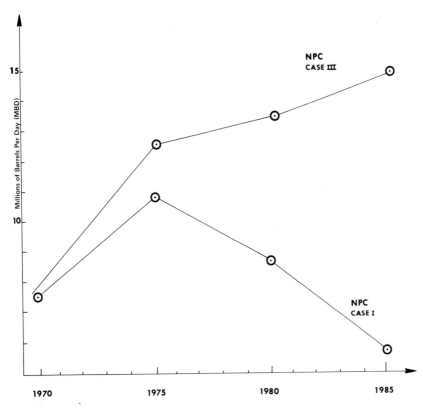

Figure 1-1. Demand For Foreign Waterborne Petroleum, NPC Study.
Source: Table 1-11.

The present international trend toward the use of larger and larger tankers for petroleum transportation is the natural result of the entrepreneur's desire to utilize economies of scale on longer trips. Yet the huge tankers produced for the international market by these economies of scale are simply too large to be accommodated by the conventional ports of the East and Gulf coasts. The draft of some supertankers actually exceeds the depth of water of all conventional United States ports. As discussed earlier in this chapter, tanker draft is both a physical and an economic barrier to entry.

To understand the trend toward large tankers one must understand the tanker market structure. The evidence clearly indicates declining average cost for delivery of oil as tanker size is increased. Petroleum coming from the Middle East and other foreign markets comes to the United States in tankers that are part of the international market structure. The international tanker market is differentiated from the United States coastal market by the Jones Act, which requires that cargoes shipped between ports in the United States be carried in

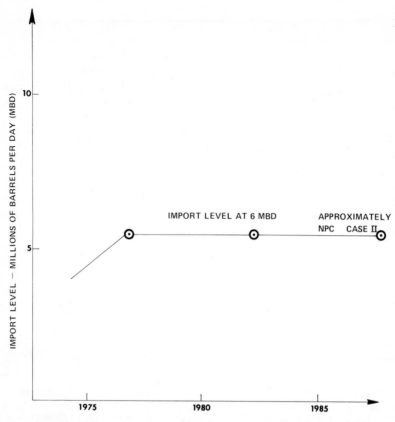

Figure 1-2. A Forecast of Petroleum Imports. Source: Table 1-11.

American ships.[h] If the quantities of oil indicated in Figure 1-2 are to be shipped to the United States, the scale economies of international tanker systems and markets must be explored.

Zenon Zannetos in a definitive study examined the nature of international tanker markets. He concluded that, although "on the basis of common sense and a priori characteristics [the markets] should be imperfect," in fact they operate like "perfectly competitive markets."[20] This conclusion is derived by a consideration of the basic elements of a competitive industry and by empirical data concerning price behavior and capacity changes over time.

There are four classic elements of proof indicating a perfectly competitive market. First is the existence of a large number of competitors, none large

[h]The Merchant Marine Act of 1920, 41 Stat. 988, declared it be "the policy of the United States to do whatever may be necessary to develop and encourage the maintenance of . . . a merchant marine."

enough to influence the market price. Zannetos counted over six hundred tanker owners, none exerting control over more than 7 percent of capacity. Morris Adelman, in his thorough study *The World Petroleum Market*, reported that by 1972 the eight largest oil companies owned only a fifth of tanker capacity, a concentration in decline since World War II. He also examined the varying duration of charters, determining that about one third of aggregate capacity comes available for charter in a given year.[21] Therefore, at any one time there are a large number of competitors and a marked absence of concentration in the marketplace.

The second element of proof concerns homogeneity of product. While some design variations exist, tankers in an economic sense are containers and undifferentiated as they add value in time and space utility to crude and refined product. The third element is ease of entry and exit from the market. There does not appear to be a significant competitive advantage (because of lower operating costs) in owning a large number of tankers. In fact, Adelman points out that many operators become owners only "to take advantage of a particular opportunity to buy one or a very few ships. But to say that many competent firms cluster on the boundaries of the industry, and that minimum capital requirements are low, is to say that entry is easy and market control impossible."[22] This is clear definition of absence of barriers to entry.

The fourth and final element of proof of perfect competition is having all economic agents possess complete and perfect knowledge of market transactions. This is essentially the case in the international tanker market, as chartering is brokered in centrally located markets in New York, London, and Oslo, and by the wide use of standard rate schedules (i.e., worldscale,[i] permitting price comparison between charters).

Price data for the industry strongly support the notion that is basically competitive. In the short run the single voyage or "spot" rate is determined by aggregate supply and demand, depicted in Figure 1-3. The spot rate fluctuates as supply and demand shift. Owners are price takers but can adjust to market conditions to the extent that system buffers—such as maintenance availabilities, voyage speed adjustments, and turnaround times—will permit. Price expectations ought to be important in the decision to increase supply (i.e., build more tankers), and presumably owners would expect rising prices. In this case a simple dynamic equilibrium can exist, with prices rising over time. However, should there be some exogenous shock to the expected demand function, prices will change substantially. Suppliers will attempt to adjust to the new market price, but at the same time capacity ordered earlier anticipating higher prices may enter the market, shifting the supply function to the right and allowing the construction gestation lag to further reinforce the pressure of demand shift. This

---

[i]Worldscale is a schedule of tanker rates based on a hypothetical ship. Issued by the International Tanker Nominal Freight Scale Association, Ltd., London, Worldscale has been in effect since September 1969.

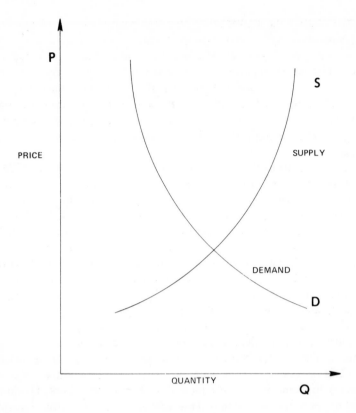

Figure 1-3. Short Run Determination of Tanker Rates in the Spot Market

increase in supply can be aggravated further by expirations in "time charters." (A time charter is for an extended period and may last twenty years.) The lag in the adjustment processes combined with relatively inelastic tanker demand and elastic supply creates a cobweblike effect of widely fluctuating prices. Figure 1-4 shows just such a fluctuation pattern in the spot market rates of recent years.[j] This series is presented as Figure 1-4.[23] Spot rates fell sharply after the Middle East War began in October 1973. In the months preceding the war, the index of charter rates skipped from worldscale 350 to 420, and on one day in October 1973 dropped from 420 to 80. The index rebounded to about 200, cascaded back to the 80 level, and fell still lower in 1975.[24]

In a competitive market capable of such price fluctuation, some tankers will,

[j]The spot rate is set for a single voyage; see Morris Adelman, *The World Petroleum Market* (Baltimore, Md.: Johns Hopkins University Press, 1972), p. 103, for an excellent discussion of spot rates.

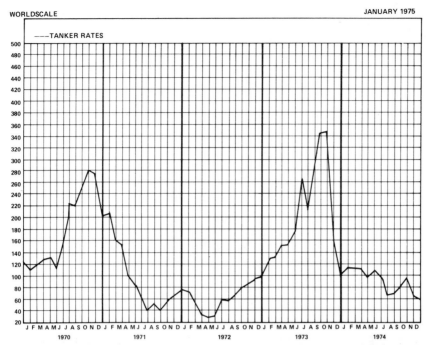

Figure 1-4. Single Voyage Worldscale Tanker Rates (Base Persian Gulf to Western Europe) for Tankers between 40,000 and 99,999 DWCT. Source: *Shipping Statistics & Economics*, H.P. Drewry (Shipping Consultants) Limited, London, England. January 1975.

in the short run, not be able to cover costs. As market prices fall, owners as price takers continue to enjoy profit as long as spot rates are above the "continuation rate." (The continuation rate is loosely defined as the rate at which an owner would continue to sail his ship. Often an owner might sail a ship at a rate below cover of variable cost, just to avoid layup and reactivation costs.) Otherwise they struggle along in the transfer rate zone, or lay up their ship as spot or limited voyage charter rates fail to cover variable costs. This is pictured in Figure 1-5. The layup decision has been discussed by Sturmey, Adelman, and Zannetos; obviously it always involves a myriad of factors, including the costs of layup and of reactivating a ship at a later but more opportune time. After the 1974-5 tanker market, layups exceeded 10 percent worldwide.[25]

Considerable speculation can exist between the short-run spot market and the long-run time market; such speculation occurred in the 1973-75 market gyrations.[26] At any market rate below the long-run average cost, short-run circumstances may allow profitable operation; however, the number of new ships on order will drop sharply. This happened in 1974-75.

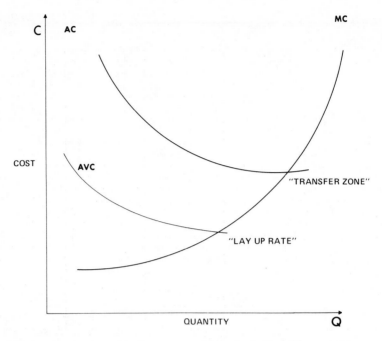

Figure 1-5. The Lay Up Decision and Marginal, Variable, and Average Costs

In the long run the market senses that the equilibrium rate is the time rate. This rate is dependent on the construction and operation costs over the period for the "incremental ship," i.e., the present discount value of the earnings of the incremental ship would equal the cost of construction. Given these costs and the interest rate expectations, the necessary time rates for tankers are automatically determined.[27] Adelman describes how the time rate can serve as an anchor for the spot, as shown in Figure 1-6. Time rates have been higher than spot rates because of the shortage of efficient new tankers, but Adelman notes that when "the shortage eases, ship prices and long term rates could ease."[28] In this fashion the time-rate anchor can drag if prices fall and trigger the dynamic cobweb pattern described earlier. The 1974 market may again be an example.

Tanker owners in the international market have all the problems of a price taker in a competitive market. They must be concerned with marginal cost structures, technological and regulatory innovation in construction and operation, and any other cost-saving promising a downward shift in costs. Understanding this concern is essential in fathoming the trends in tanker construction. The acceleration in size in recent years has come from a multitude of forces; by elastic price expectations, by shipbuilding innovation in Japan and elsewhere, by

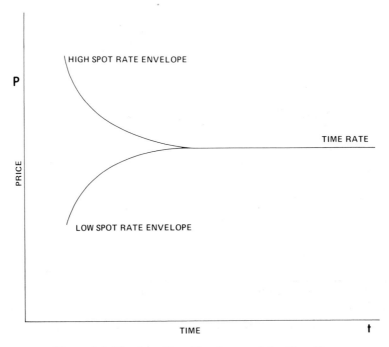

Figure 1-6. The Spot Rate Envelope and the Time Rate

design innovation, by scale economies in crew and technical systems, and not in the smallest measure by learning curve phenomena made possible by the expansion of world petroleum trade driven by worldwide increases in energy demand. A typical declining cost curve for oil transport on a long international voyage is indicative of declining average cost by increasing returns to scale and elastic productivity. (Specific average cost curves will be presented and discussed in Chapter 3.)

Tanker owners by competitive necessity have had to focus on classic economic variables of cost and capacity. Since 1960 this has resulted in the acceleration in size of newly constructed tankers, and in their use as operating systems with appropriate terminal facilities. Tanker growth has become so rapid that the "world's largest tanker" has been nine different ships in sixteen years. In this period the largest tanker has grown from 56,000 to 477,000 tons carrying capacity. This is a tenfold growth in ship size; if tanker size were only to double from here the world would have a "megatanker." Figure 1-7 arrays the dimensions of a T-2, the *Idemitsu Maru*, the *Nisseki Maru*, and the *Globtik Tokyo*. Where the T-2 had a draft of only 30 feet, the "superships" have drafts in excess of 60 feet.[29] ("Supership" is the term coined by Noel Mostert, in his book of the same name published in 1974 by Knopf.)

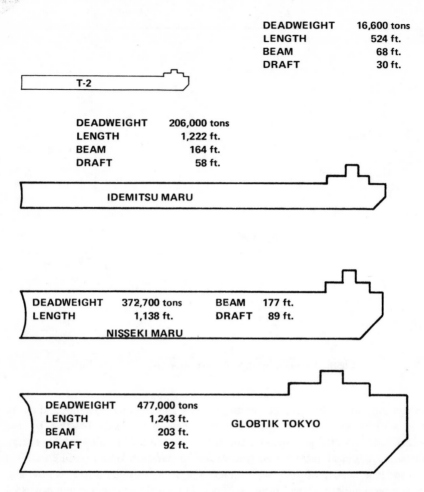

DEADWEIGHT	16,600 tons
LENGTH	524 ft.
BEAM	68 ft.
DRAFT	30 ft.

T-2

DEADWEIGHT	206,000 tons
LENGTH	1,222 ft.
BEAM	164 ft.
DRAFT	58 ft.

IDEMITSU MARU

DEADWEIGHT	372,700 tons	BEAM	177 ft.
LENGTH	1,138 ft.	DRAFT	89 ft.

NISSEKI MARU

DEADWEIGHT	477,000 tons
LENGTH	1,243 ft.
BEAM	203 ft.
DRAFT	92 ft.

GLOBTIK TOKYO

Figure 1-7. The Dimensions of Some Supertankers. Source: *Technology Review*, Volume 75, No. 5 (March/April, 1973), p. 57.

At the start of 1974 there were 3293 ships of 212 million deadweight tons in the world tanker fleet. These tankers fly the flags of many nations: Liberia, Japan, Britain, Norway, and Greece, respectively, have the largest number registered. The size of the average tanker in the world fleet has increased from forty thousand tons in 1970 to the mid-fifty-thousand-ton range by 1975.[30] The average size statistic is increasing rapidly as smaller tankers, comprising a large percentage of the older tanker fleet, leave the fleet inventory as they reach the end of their economic life. At the same time newly constructed tankers entering the inventory are substantially larger than average size, also causing an increase in the fleet average size.

Before the precipitous decline in the world tanker charter market following the 1973-74 embargo, over twelve hundred tankers were on order from world shipyards. The aggregate tonnage represented by these orders was nearly two hundred million tons; delivery of these tankers would radically alter existing tanker size distributions. Table 1-12 presents Fearnley and Egers data for 1974 on flag and size of tankers in operation; Table 1-13 presents their data on size and location of tanker construction on order at that time.[31] If the demand for petroleum transportation decreases over the next twenty years (the length of many term charters for new construction), aggregate new tonnage on order may continue to decline, as it has since the 1973-74 oil embargo. A period of rapidly changing and generally low charter rates can be particularly difficult for the independent tanker owner. For example, in 1973 the International Association of Independent Tanker Owners, or Intertanko (comprising approximately 70 percent of independent tank owners), found that "special measures were

Table 1-12

The Tonnage Range and Country of Registry of Tankers in Operation, 1974

Tonnage Range (Deadweight Tons)	Percentage	Registry Flag (Country)	Percentage
10-60,000	30.9	Liberia	28.1
60-100,000	16.4	Britain	12.4
100-150,000	8.6	Japan	12.4
150-200,000	3.2	Norway	9.8
200-250,000	26.4	Greece	5.7
250-300,000	12.3	Remaining	31.6
300,000 and over	2.2		

Source: C. Hayman, "What to Do With All Those Tankers," *New York Sunday Times*, April 14, 1974, p. F-1, in part from Fearnley and Egers Chartering Company, Ltd.

Table 1-13

The Tonnage and Location of Tanker Construction, 1974

Tonnage Range (Deadweight Tons)	Percentage	Construction Location (Country)	Percentage
10-60,000	4.1	Japan	51.7
60-100,000	6.7	Sweden	9.6
100-150,000	12.5	West Germany	5.7
150-200,000	3.9	Spain	5.2
200-300,000	42.3	Britain	4.5
300-400,000	17.3	Remaining	23.3
400,000 and over	13.2		

Source: C. Hayman, "What to Do With All Those Tankers," *New York Sunday Times*, April 14, 1974, p. F-1, in part from Fearnley and Egers Chartering Company, Ltd.

required to rescue those owners trapped into long-term charter agreements" set at rates that have since been substantially altered by currency floats and revaluations.[32] A substantial decrease in tanker demand will cause orders for new construction to be canceled, and this in turn will affect the charter market, and average tanker size.

Tankers and Deepwater Terminals as Operating Systems

The terminals that exist to load, unload, and service a tanker after each leg of a voyage determine, by their design and effectiveness, the turnaround time for the tanker. Were it not for "external diseconomies" in filling and discharging oil, increasing returns to scale in tanker construction and operation that now exist indicate that the optimal size of a tanker is infinite.[33] Under these conditions, the optimal way to move an infinite volume of oil is a pipeline. Therefore, at least conceptually, optimal tanker size is a "pipe."[k] Optimal size is obviously not infinite if a probable voyage the tanker must undertake requires it to sail to a port where the tanker draft will cause it to go aground miles before it reaches port.

Optimal tanker size might not be infinite if the expected environmental costs of oil spills, when added to total petroleum transportation costs, effectively cause average cost beyond a certain tanker size to rise. Work done on this subject implies strongly that this is not the case: environmental costs are not a barrier to tanker size, per se.[34] (This subject will be pursued in Chapter 4.)

In the United States port size presents a formidable barrier to tanker size. The 206,000-, 372,000-, and 477,000-ton tankers presented in Figure 1-7 have drafts of 58, 89, and 92 feet. These tankers could not be serviced in any United States port. Long Beach, California could handle a tanker of 100,000 deadweight tons; this is the largest port in the "lower 48." Stated bluntly, large ships that have appeared in response to the scale economies and demands of an international market are too large to fit in conventional American ports. Innovation, specifically with regard to deepwater terminals, will be required if the United States is to take advantage of the scale economies of the world tanker system. Existing port size and deepwater terminals for the United States will be discussed in detail in succeeding chapters.

A thorough grasp of the tanker market is essential in understanding the workings of the marine petroleum transportation system. If the United States is to depend from 1975 to 1985 on imported petroleum in the magnitude

---

[k]The pipe concept is discussed in Louis K. Bragaw, "Environmental Policy Formulation in Competitive Tanker Markets, *Decision Sciences Northeast Proceedings*, Vol. 3 (1974), pp. 31-34. The idea was suggested by Robert N. Stearns of Connecticut College, while he critiqued a draft of the article, and pursued by Irving H. King of the United States Coast Guard Academy in many discussions.

forecasted in Figure 1-2 (p. 16), approximately 6 MBD, or any higher level, both supertankers and deepwater terminals must be considered as operating systems. The marine petroleum transportation system, conceptually at least, is a "pipe," and as such should be examined from the perspectives of economics and the environment, to see that as an operating system it best serves the nation.

2

The Location of Deepwater Terminals

The Trend Toward Deepwater Terminals

How should the United States respond to the trend toward development of marine petroleum transportation systems consisting of supertankers and connecting deepwater terminals? Are not these systems really "supersystems," conceptually at least the "pipe" discussed in Chapter 1 as the ultimate result of continuously increasing scale returns? Is it really necessary for the world's largest industrial trading nation and consumer of energy to have the ability to receive supersized tankers? What would be the consequences if deepwater terminals were not constructed to interface with superships, and complete the emerging marine petroleum transportation system? The following chapters will try to examine these questions from an overall view, focusing on economic, environmental, political, manpower, and technological considerations.

Chapter 1 emphasized dimensions of the geographic imbalance between aggregate petroleum supply and demand. A matrix of forecasts of petroleum imports necessary to satisfy that imbalance for the United States for the years 1975, 1980, and 1985 was developed. The nature of the tanker transportation cost function and the chartering market was discussed; the resulting acceleration in absolute tanker size since the mid-1960s was traced. Supertankers built since then simply are too large to enter most United States ports. The most significant physical barriers to entry for large, fully laden tankers are the depths (and widths in some cases) of domestic harbors and entrance channels. Economic and environmental barriers to entry include social and economic costs of waterfront land, problems in dredging, and environmental damage possible through oil spills from collisions or groundings in congested inner habors.

United States harbors and entrance channels are grossly undersized for vessels over 100,000 deadweight tons. The Port of Long Beach is deepening its main ship channel to 62 feet at mean low water; it will be the only United States port capable of unloading a 200,000-ton tanker at berth.[1] As one can see from examination of Table 2-1, the East and Gulf coasts of the United States have no comparable facility. Portland, Maine, and Beaumont, Texas, can handle 80,000-ton tankers; after these ports East and Gulf coast terminal capacities drop rapidly. The majority of the ports are only deep enough in their main ship channels and berthing facilities to accept vessels of 35- to 40-foot draft. This represents a tanker in the 30,000-to-50,000-deadweight-ton capacity range.[2] Figure 2-1 presents the relationship between design draft and tonnage. Table 2-1 arrays the constraining depth limit of domestic tanker ports.

Table 2-1
Port and Harbor Capabilities

Port of Harbor Area	Max. Draft Vessel Using Area (feet) 1969	1970	Controlling Depth (feet)	Aver. Tidal Variation (feet)	Approx. Max. Dredgable Depth (feet)
Portland, Me.	51	51	45	9.0	60.
Boston	41	42	40	9.0	40/60
New York (Ambrose)	45	44	45	4.5	60(Narrows)
(Kill Van Kull)	40	40	35	5.0	38
Delaware Bay to Philadelphia	46	46	40	4.1 to 5.9	41
Philadelphia, Pa.	40	41	40	6.0	41
Baltimore, Md.	40	40	42	1.2	60
Hampton Roads, Va.	44	47	45	2.5	60
Jacksonville, Fla.	35	35	40	2.5	44
Port Everglades, Fla.	38	39	40	2.5	42
Tampa, Fla.	35	35	34	2.0	>40
Mobile, Ala.	40	40	40	1.5	45
Pascagoula, Miss.	38	39	38	1.6	50
New Orleans, La.	40	39	40	1.0	45
Baton Rouge, La.	40	40	40	1.0	40
Beaumont, Tex.	37	39	40	1.0	47
Galveston, Tex.	39	40	36	2.0	52
Houston Ship Channel	40	40	40	1.0	45
Corpus Christi, Tex.	40	39	45	1.0	50
Los Angeles, Calif.	52	45	51	5.4	>80
Long Beach, Calif.	46	51	52	5.5	>80
San Francisco	50	51	50	5.7	>100
Bay Entrance, Calif.					50 (Interior)
Columbia River	38	38	42	2.0	48
Entrance Puget Sound	39	39	100-500	1.1	>100

Source: Waterborne Commerce of the U.S., Corps of Engineers, U.S. Army.

Ports of the East and Gulf coasts were world leaders for the largest tankers in service in 1960. Japanese and European ports could also service 35,000-to-40,000-ton tankers, the largest ships at the time. Since then tanker scale economies and demand increases in the international petroleum transportation market have brought into being a vastly changed marine petroleum transportation system. The supertanker construction boom since 1960 led to deepwater terminals necessary to service them and minimize total transportation costs.

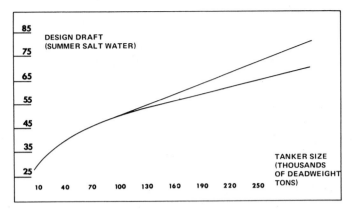

Figure 2-1. Design Draft of Supertankers and Deadweight Tonnage. Source: *Technology Review*, Volume 75, No. 5 (March/April, 1973), p. 57.

By 1974, the United States became virtually surrounded by nations with deepwater terminal facilities capable of servicing supertankers. In those foreign ports where adequate natural harbor and channel depths were not available or easily constructed, transfer terminals were designed and installed several miles offshore to attain the required deep water. The most common terminal facility is the single buoy mooring or single-point mooring; there have been over one hundred installations, many of them capable of handling the largest supertankers in operation at the time of installing the terminal.[3] The single buoy mooring and other terminal options will be discussed in detail in the next chapter.

In many countries, dealing in the crude and refined product trade, the guiding philosophy for such marine petroleum transportation systems is that the port that expands the fastest will get the business of the future. In the United States, environmental and other social and political concerns have been weighed before deepwater terminals have been established. It is obvious that other systems worldwide will by substitution fill the void if deepwater terminals are not constructed in the United States. Such substitution would mean not only increased costs to consumers because of transshipment in smaller tankers, and possibly increased environmental hazards, but also loss of jobs in the industry. Moreover the United States would have no control over such deepwater ports in foreign countries, and would suffer balance-of-payment losses.

Chapter 1 discussed the supertanker and deepwater terminal as an operating system for marine petroleum transportation. While the supertanker is a super-ship, the deepwater terminal is really a superterminal. Conceptually they are the "pipe"—resulting from increasing system scale returns—for marine petroleum transportation; as such they form a "supersystem." If such a supersystem is necessary, one must examine it in detail to assess its impact. This entails examining aggregate demand for petroleum imports generally; specifically, a regional analysis of petroleum shortages should be conducted to determine which geographic areas are most likely to receive imports.

Analysis of Regional Petroleum Shortages

Several methods can be employed to forecast the regional imbalances between petroleum supply and demand. An aggregate technique will be employed to illustrate some of the factors that will be involved in the location decisions that will be made. This technique will provide approximate values only. The analysis will focus on projections for 1985, using various available data and making several assumptions.

First, demand patterns should be examined. Case II developed by the National Petroleum Council, discussed in Chapter 1, estimated petroleum demand for 1985 to be 23 million barrels per day (MBD). Regional demand patterns should also be studied. For administration of national security and oil import programs, the United States is divided into five Petroleum Administration for Defense (PAD) Districts, as depicted in Figure 2-2. The districts, or markets, when viewed separately, exhibit a markedly different quantity and mix

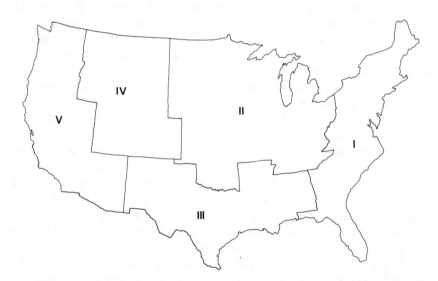

Figure 2-2. The Five Petroleum Administration for Defense (PAD) Districts

of crude and product demand. From Figure 2-2 it can be seen that PAD I covers the states along the East Coast (including the Florida Gulf Coast), PAD III covers the Gulf Coast, PAD V the West Coast, etc. Table 2-2 arrays regional petroleum demands, by PAD, for 1970. These figures are from National Petroleum Council reports; the use of their figures in the analysis that follows is, as it was in Chapter 1, not an endorsement of these figures, but a reflection of their similarity with recently published information. If one assumes that each PAD district maintains the same percentage of total demand in 1985 as it did in 1970, petroleum demand by PAD in MBD can be easily projected (they are arrayed in the right hand column of Table 2-2).

If the United States achieves self-sufficiency in refinery capacity by 1985, and if the 1973 refinery capacities as reported by the American Petroleum Institute are accepted, then the gap (or excess) in refinery capacity by PAD can be calculated. Table 2-3 provides the figures for determining needed refinery capacity or final product imports. If one assumes that refineries can be built by 1985 to meet this 9.5 MBD increase in petroleum demand, sources of crude to supply the refineries will still be needed. In addition to present domestic production, this crude may come from the three main supply options, i.e., the Trans-Alaska Pipeline System (TAPS), the Outer Continental Shelf operations (OCS), or from imported petroleum. Still other options may come from new technologies. This analysis will ignore the possibility of decreases in demand; it will assume any such decreases will reduce the amount of petroleum imported.

Domestic petroleum production for 1985 has been estimated at 9.5 MBD. The throughput available from TAPS can be estimated at 2.0 MBD.[4] Various estimates have been given for the energy value of new technologies that may be available by 1985; obviously, this is a very complex issue. For purposes of illustration in this analysis, a new technology supply of 0.75 MBD will be assumed. This estimate could come, for example, from shale oil; estimates in this

Table 2-2
Petroleum Demands by PAD for 1970 and 1985

PAD	1970 Demand MBD	Percent of Total	1985 Demand MBD
I	5.966	40	9.20
II	4.037	27	6.21
III	2.537	17	3.91
IV	0.367	2	0.46
V	2.070	14	3.22
Total	14.977	100	23.00

Source: *U.S. Energy Outlook: An Initial Appraisal, 1971-1985; An Interim Report*, National Petroleum Council, 1971, Vol. 2, pp. xv and 59.

Table 2-3
Projections of Refinery Capacity Addition by 1985

PAD	1973 Capacity MBD	1985 Demand MBD	1985 New Refinery Requirements MBD
I	1.809	9.20	7.391
II	3.537	6.21	2.673
III	5.522	3.91	(1.612)
IV	0.432	0.46	0.028
V	2.200	3.22	1.019
Total	13.500	23.00	9.500

Source: American Petroleum Institute, U.S. Refinery Operations (Washington, D.C., April 10, 1973), p. IV-3.

range have been made for shale by 1985.[5] If demand for 1985 is 23.0 MBD, this means that 10.75 MBD must be supplied by either petroleum imports or OCS operations. Table 2-4 arrays these estimates.

If it is further assumed that each PAD will maintain oil production in the same percentage of total production in 1985 that it produced in 1970, and the estimate of 9.5 MBD production for 1985 is utilized, domestic crude production for 1985 can be calculated. These figures are presented in Table 2-5. Each PAD can now be examined by aggregate analysis to balance supply with demand for 1985. A district by district analysis will be followed.

When tradeoffs are made, between constructing a refinery or importing

Table 2-4
Estimates of Petroleum Sources in 1985

Source	Quantity-MBD
Domestic petroleum	9.50
TAPS	2.00
Shale[a]	0.75
OCS and Imports	10.75
Total	23.00

[a]Shale has been used to illustrate how a new "source" can be handled in this type of analysis. Both the absolute amount and the geographical location of the new source are obviously critical to the analysis. Shale as a new source appears to have many problems; it most probably will not be one of the first new sources to appear in the 1980s.

Sources: "Guide to the National Petroleum Council Report on United States Energy Outlook" (Presentation Made to National Petroleum Council, Washington, D.C., December 11, 1972), pp. 6-22; and *U.S. Energy Outlook*, A Report to the National Petroleum Council's Committee on U.S. Energy Outlook, December 1972, pp. 272-3.

Table 2-5
Domestic Petroleum Production in 1970 and Projections for 1985

PAD	1970 MBD	Percentage	1985 MBD
I	0.057	0.5	0.048
II	1.433	12.6	1.197
III	7.812	69.0	6.555
IV	0.708	6.3	0.598
V	1.318	11.6	1.102
Total	11.328	100.0	9.500

Source: National Petroleum Council, see Table 2-4.

refined product into a PAD, consideration must be given to the following: availability of crude, quantity of crude, transportation mode, and total cost of refined product. For this reason the aggregate analysis will attempt to assign refined product produced in a PAD to consumption within the PAD. If a surplus of refined product exists in a PAD, it obviously must be exported to another PAD. Using the data discussed here or appearing in tables, and following the general criteria just discussed, the districts in need of increased refinery capacity can be determined. The data necessary for this analysis is presented in Tables 2-6 and 2-7; Figure 2-3 graphically displays how aggregate analyses of this type can be conducted.

Now that the location of increased refinery capacity for 1985 has been determined by PAD, the new refinery capacity can be located by such determinants as environmental concerns, market demand, crude availability, labor costs, and transportation facilities. Although the analysis used here to locate refineries is not detailed, great consideration was given to integration of

Table 2-6
Aggregate Supply Mix by PAD for 1985

PAD	Domestic Production (Table 2-5) MBD	Imports and OCS Supply MBD	Shale Supply (Table 2-4) MBD	TAPS Supply (Table 2-4) MBD	Total Crude Supply MBD
I	.048	9.152	0	0	9.200
II	1.197	1.480	0	0	2.677
III	6.555	0	0	0	6.555
IV	.598	0	.750	0	1.348
V	1.102	.118	0	2.000	3.220
Total	9.500	10.750	.750	2.000	23.000

Table 2-7
Locating New Refinery Capacity by Aggregate Analysis, and PAD Transfer

PAD	1973 Existing Ref. Capacity (Table 2-3) MBD	1985 Projected Ref. Capacity (Table 2-3) MBD	1985 Projected Demand (Table 2-2) MBD	1985 PAD Transfer Export or (Import) MBD
I	1.809	7.391	9.200	0
II	3.537	.340	6.210	(2.334)
III	5.522	0	3.910	1.611
IV	0.432	0.750	0.460	.722
V	2.200	1.019	3.220	0
Total	13.500	9.500	23.000	0

the new facilities with existing supply networks. Table 2-7 shows the distribution of new refinery capacity forecast for 1985 by the preceding aggregate analysis. Table 2-7 underscores the fact that, using this analysis and the numbers that have been assumed, nearly all of the 10.75 MBD import or OCS supply expected is needed on the East and Gulf coasts of the United States. These figures take into account assumptions of TAPS delivery of 2.0 MBD to PAD V,

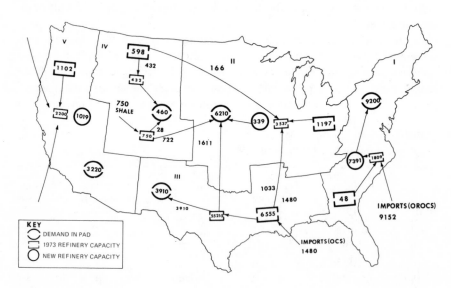

Figure 2-3. Allocating Petroleum Supply to Accommodate Demand Projections for 1985, by Aggregate Analysis over PAD Districts

and a new technology source from shale of 0.75 MBD in PAD IV. It appears strongly that new supply in the form of imports is needed on the East and Gulf coasts.

Tanker Traffic and the Need
for Terminals

The amount of petroleum that will be imported to the United States can serve as an input to a simple model to estimate the number of tankers that will be required to deliver the petroleum. The purpose of the model is only to illustrate possible tanker traffic patterns and voyage frequencies, thereby demonstrating another aspect of the need for deepwater terminals. The model assumes that tankers operate 350 days a year, have a speed of 15 knots (nautical miles per hour), and spend 4 days per voyage on turnaround and contingencies. The model assumes that nearly all of the imports will come to the East and Gulf coasts, while a negligible portion will be delivered to the West Coast. Specifically, it will be assumed that imports are divided as follows: from the Persian Gulf, 5 percent to PAD V, 45 percent to PAD I, and 35 percent to PAD III; from the Mediterranean, 15 percent to PAD I.

These assumptions must be supplemented with a size of tanker to be used; calculations are made for tankers of 60,000, 100,000, and 250,000 deadweight (carrying capacity) tons. Table 2-8 lists the cargo-carrying capacity of tankers of these sizes for the various routes and previously discussed assumptions. Given the number of round-trip voyages that a tanker can make in a year and its cargo capacity, the annual cargo capacity for the particular tanker size can be calculated. Table 2-9 presents these figures. Given the annual capacity of tankers making these voyages, the numbers of tankers that will be required to transport

Table 2-8
Cargo-Carrying Capacity for Tankers on Selected Routes

Route	Voyages/Year	Cargo Capacity/Tanker (In Tons)		
		60,000	100,000	250,000
Persian Gulf to PAD V and return	5.2	58,200	97,600	245,600
Persian Gulf to PAD I and return	5.0	57,800	97,200	244,700
Persian Gulf to PAD III and return	4.9	57,800	97,000	244,500
Mediterranean to PAD I and return	12.0	58,700	98,300	246,800

Table 2-9
Annual Cargo Capacity for Tankers on Selected Routes

	Annual Capacity for Tankers of Size: (in Tons)		
Route	60,000	100,000	250,000
Persian Gulf to PAD V and return	302,640	507,520	1,277,120
Persian Gulf to PAD I and return	289,000	486,000	1,222,500
Persian Gulf to PAD III and return	283,220	475,300	1,198,050
Mediterranean to PAD I and return	704,400	1,179,600	2,961,600

a given level of petroleum imports can now be calculated. Table 2-10 displays such calculations for assumed import levels of 2 million barrels per day (MBD); a mix of 100,000- and 250,000-ton tankers could be used, but the calculations determine the number of either 100,000- or 250,000-ton tankers. Table 2-11 expands the preceding table by presenting the number of either 100,000- or 250,000-ton tankers that would be required at import levels of 2, 6, and 12 MBD. Obviously, the 6 and 12 MBD levels are multiples of the 2 MBD level. Tanker requirements for a 10.75 MBD import level (i.e., the import level estimated for 1985 in Table 2-4) are tabulated in Table 2-12. Table 2-13 tabulates tanker requirements on each of the four previously discussed voyages, assuming that the entire cargo is carried in either 100,000-ton or 250,000-ton tankers.

Table 2-10
Estimates of Number of Tankers Needed to Deliver an Import Level of 2MBD

Route	Imports MBD	Imports Million Tons/Year	Tankers Required[a] Just 100,000	Just 250,000
Persian Gulf to PAD V and return	.1	5	10	4
Persian Gulf to PAD I and return	.9	45	93	37
Persian Gulf to PAD III and return	.7	35	74	29
Mediterranean to PAD I and return	.3	15	13	5
Total	2.0	100	190	75

[a]Number of tankers are rounded to nearest integer.

Table 2-11

Total Number of Either 100,000- or 250,000-DWT Tankers Required to Carry Varying Levels of Imported Petroleum

	Number of Tankers Required for the Following Level of Petroleum Imports[a]		
Size Tanker Used	2 MBD	6 MBD	12 MBD
100,000	190	567	1133
250,000	75	225	450

[a]Numbers of tankers rounded to nearest integer.

Table 2-12

Total Number of Either 100,000- or 250,000-DWT Tankers Required to Carry a Forecasted Import Level of 10.75 MBD

Size Tankers Used	Number of Tankers Required[a]
100,000	1015
250,000	404

[a]Number of tankers rounded to the nearest integer.

Table 2-13

Total Number of Either 100,000- or 250,000-DWT Tankers Required to Carry a Forecasted Import Level of 10.75 MBD over Four Representative Routes

Routes	All Traffic 100,000 DWT[a]	All Traffic 250,000 DWT[a]
Persian Gulf to PAD V and return	51	20
Persian Gulf to PAD I and return	456	182
Persian Gulf to PAD III and return	356	141
Mediterranean to PAD I and return	152	61
Total	1015	404

[a]Number of tankers rounded to nearest integer.

Figure 2-4 roughly depicts the various voyages assumed in this example. The numbers on the table are the number of 100,000-ton tankers that would be required to deliver imported petroleum at the 10.75 MBD level on the four routes. One thousand and fifteen tankers of 100,000-ton capacity constitute considerable traffic; they certainly demand serious consideration for deepwater terminals.

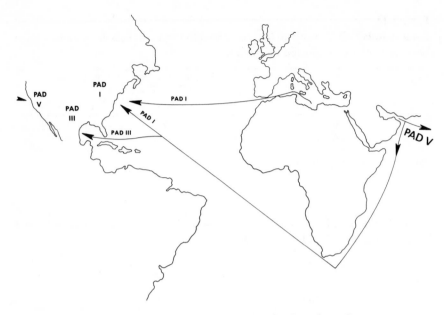

Figure 2-4. Illustrative Tanker Tracklines for Petroleum Imports

Locating Deepwater Terminal Facilities

Deepwater terminal locations could be decided by a multitude of means. The aggregate analysis by PAD discussed earlier highlights the necessity from an economic viewpoint for deepwater terminals on the East Coast of the United States. It would be natural to ship crude to areas of heavy refinery concentration, and refined product to areas of heavy final product demand. The Texas Gulf Coast contains the heaviest concentration of oil refining and petrochemical processing industries in the world. The capacity of these refineries and the established position of the petroleum industry in the Gulf Coast region, PAD III, make the region a natural site for locating deepwater terminals, for no other reason than integrating existing refinery facilities with the supersystem. The second largest concentration of refineries in the United States—more than 90 percent of East Coast capacity—is located at the Delaware River estuary. From an economic viewpoint, it is obvious that deepwater terminals should be on the East and Gulf coasts of the United States. The problem that remains is selecting specific locations.

One type of analysis that could be used would proceed in general as follows. Demand for petroleum could be modeled for geographic areas as a function of past consumption and population projections for each statistical metropolitan area and contiguous rural countries. At the same time a supply case could be built by ascertaining the capacity of production, transportation of crude, refinery, and final product distribution capabilities in a given area. The sum of the supply systems for areas under consideration could then—given input assumptions—heuristically feed the demands generated by the models developed for successive geographic areas. The marginal source of supply to meet demand would be imported petroleum. Minimization of total economic cost would then be used as an objective function, along with other considerations, to determine optimal areas for deepwater terminals. A technique similar to this, although more complex, was developed in an excellent study prepared for the U.S. Department of Transportation.[6]

The Department of Transportation model utilizes inputs of refinery supply and area demands with "enough imported product to make supply and demand equal." Essentially, their model has each of twelve refinery districts supply the nearest demand area, and in successive iterations, the successive nearest demand area, etc. The model examines several potential deepwater port areas and area combinations, seeking least-cost transportation combinations.

With the tremendous concentration of industrial activity along the East and Gulf coasts massive volumes of petroleum will have to be transported there whether deepwater terminals are established or not. The demand forecasts for petroleum imports discussed in Chapter 1 did not explore the regional need for imported petroleum to satisfy regional imbalances between domestic supply and existing demand patterns. The aggregate analysis by PAD described the shortage along the East Coast, and models such as the Department of Transportation model could play a central role in determining deepwater terminal location. Such location must also consider environmental, political, social, and technological factors. The next chapter will focus on the economic aspects of deepwater terminal systems. In the supersystem, deepwater terminals are the gateway to superships delivering imported petroleum.

3

Economic Aspects of Deepwater Terminal Systems

The Economic Incentive

There is indeed a basic economic incentive to build deepwater terminals. A single superport terminal system can result in an annual savings of millions of dollars. While these savings will accrue to the nation as a whole, the amount that will be passed on to consumers living near a deepwater terminal is uncertain and has become a source of concern relative to some proposed offshore terminals. Tables 3-1 and 3-2, taken from a government study directed by the Council of Economic Advisors, show the possible benefits of a single superport system on the East Coast and the Gulf Coast.[1] (This study was performed before the 1973-74 oil embargo.) The study analyzed costs for United States and foreign flag tankers both with and without double bottoms, and in addition considered a variety of potential governmental policies.

A foreign flag tanker without a double bottom is the least expensive means of ocean transport, while a United States flag tanker with a double bottom is a more expensive form. If federal legislation were passed to require that oil imported into the United States be carried in United States flag tankers with double bottoms, an economic alternative to be compared with a United States superport would be a foreign superport in the Caribbean or Canada. (The labor implications on such legislation are discussed in Chapter 7.) In this way, oil from the Persian Gulf could move to these offshore foreign superports in foreign flag tankers without double bottoms, and only the short transshipment voyage would utilize more expensive tankers of United States registry.

In all the variations considered in the Council of Economic Advisors study, VLCCs contribute more than 80 percent of the total costs of deepwater port operations. Pipeline distribution provides the least-cost means of transferring crude oil from deepwater ports to onshore facilities. The only exception to this statement occurs in the Gulf, when the daily volume is less than two million barrels. In this case, tug-barge distribution would permit slightly lower total costs.

The Council of Economic Advisors concluded that monobuoys were generally the least expensive, although there was little difference between the monobuoy type of terminal and the sea island pier in certain instances. Under extreme assumptions of the effect of weather conditions on the terminal operations, the sea island pier became the least-cost alternative at certain East Coast locations.

While the exact economic savings from a particular terminal must depend on

41

Table 3-1
Savings Resulting from an East Coast Deepwater Port

Throughput	Worst Case[a]		Best Case[b]	
(Million barrels per day)	(Cents per barrel)	(Millions of dollars per year)	(Cents per barrel)	(Millions of dollars per year)
0.600	−4.0	−8.76	3.3	7.23
0.800	−1.6	−4.67	5.7	16.64
1.000	−0.2	−0.73	7.2	26.28
1.135	0.4	1.66	7.8	32.31
1.200	1.0	4.38	8.4	36.79
1.572	3.2	18.36	10.5	60.25
2.000	5.3	38.69	12.7	92.71
2.500	6.6	60.23	14.0	127.75
3.200	7.4	86.43	14.8	172.86
5.106	8.1	150.96	15.5	288.87
6.600	9.1	219.22	16.5	397.49

[a]In this case, tankers serving U.S. deepwater ports are required to have double bottoms, while tankers serving foreign ports are not.

[b]In this case, for the most part, tankers serving both United States and foreign ports are required to have double bottoms.

Source: Congressional testimony of William A. Johnson, Energy Advisor to the Deputy Secretary of the Treasury, *Deepwater Port Act of 1973*, Joint Hearings Before the Special Joint Subcommittee on Deepwater Ports Legislation of the Committees on Commerce, Interior and Insular Affairs, and Public Works, United States Senate, Ninety-Third Congress, July 23, 1973, p. 181.

many specific details, it should be noted that volume is a key factor in such a capital-intensive venture. With sufficient volume, even the "worst" case considered will result in a profitable venture. Specifically, the economic incentive of a deepwater terminal is large; it is, conceptually at least, an extension of the "pipe" discussed in Chapter 1, the result of increasing return with increasing throughput. To examine connections between tanker and pipe, one should explore the various terminal concepts.

Terminal Concepts

A superport system consists of the terminal, supertankers, pipelines, onshore facilities, and support equipment and personnel. It is possible that existing onshore ports could be dredged to accept supertankers. One advantage to such a project would be that the port could handle all commodities, since it would not be restricted to the pipe leading to an offshore terminal. However, while the Texas ports of Galveston and Corpus Christi would like to be dredged to permit

Table 3-2
Savings Resulting from a Gulf Coast Deepwater Port

Throughput	Worst Case[a]		Best Case[b]	
(Million barrels per day)	(Cents per barrel)	(Millions of dollars per year)	(Cents per barrel)	(Millions of dollars per year)
0.600[c]	−14.2	−31.10	−4.7	−10.29
1.400[c]	−0.4	−2.04	2.7	13.80
1.805	−3.6	−23.72	3.8	25.04
2.400[c]	0.1	0.88	8.4	73.58
3.248	4.0	47.42	11.5	136.33
4.175	4.6	70.10	12.0	182.87
6.782	7.7	190.61	14.9	368.84
10.600	10.0	386.90	17.1	661.60
10.900	10.1	401.83	17.2	684.30
14.700	11.1	595.57	18.2	976.52

[a]In this case, tankers serving U.S. deepwater ports are required to have double bottoms while tankers serving foreign ports are not.

[b]In this case, for the most part, tankers serving both United States and foreign ports are required to have double bottoms.

[c]Tug-barge distribution of crude oil assumed.

Source: Congressional testimony of William A. Johnson, Energy Advisor to the Deputy Secretary of the Treasury, *Deepwater Port Act of 1973*, Joint Hearings Before the Special Joint Subcommittee on Deepwater Ports Legislation of the Committees on Commerce, Interior, and Insular Affairs, and Public Works, United States Senate, Ninety-Third Congress, July 23, 1973, p. 182.

the use of superships, it appears unlikely, for both economic and environmental reasons, that such massive dredging projects will occur.

More probably, United States deepwater ports will take the form of offshore terminals. There are many different designs for offshore terminals, some quite imaginative. One design devised by Tuned Sphere International takes the form of self-propelled sphere capable of storing one million tons of oil. A similar idea proposed by Dyckerhoff & Widmann employs a U-shaped self-propelled device also capable of storing one million tons of oil. In this case, the supertanker would moor in the middle of the U while unloading cargo. At present, it appears that neither of these innovative ideas will reach fruition in the United States. More common types of offshore terminals are conventional piers, sea islands, sea island piers, multiple buoy berths (MBBs), and single-point mooring (SPM) systems.

Conventional Piers, Sea Islands, and Sea Island Piers

Conventional piers, sea islands, and sea island piers are all platform-type structures which may be permanently attached to the bottom by pilings. Sea islands and sea island piers differ from conventional piers in their connections to shore. In the case of the conventional pier, there is an accessway from shore to the loading platform so that pipelines are supported above the water by a pipe trestle, landfill, etc. Sea islands and sea island piers are located offshore and use submarine pipelines to transfer cargo to shore. A sea island pier basically resembles a conventional pier or a small complex of conventional piers located offshore, while a sea island can consist of hundreds of acres of filled land with cargo storage facilities. Such an artificial island complex, as pictured in Figure 3-1, could transfer its cargo ashore with shuttle vessels in lieu of submarine pipelines.

Conventional piers, sea islands, and sea island piers all require tugs for mooring large tankers. Launches are sometimes also required. In all cases, the tankers are moored with a fixed heading by mooring lines connected to mooring dolphins or shore mooring points.

The environmental conditions affecting the operation of ships at conventional piers, sea islands, and sea island piers are the same for these three concepts. Vessels can moor in eight-to-ten-foot-head seas or three-to-four-foot-beam seas even in winds as high as fifty knots. Cargo transfer operations, however, are normally discontinued if winds are more than thirty knots. Since tugs must assist tankers during berthing at these facilities, berthing conditions are typically limited to three-to-four-foot seas and twenty-five-knot winds.

Conventional piers and sea islands have the advantage that they can conven-

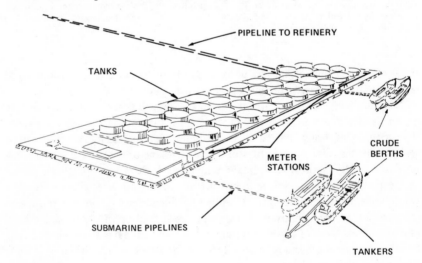

Figure 3-1. An Artificial Island Complex. Source: *Volume I, Draft Environmental Impact Statement, Maritime Administration Tanker Construction Program*, N.T.I.S. Report No. EIS 730392 D, 1973, p. IV-210.

iently handle several products or cargoes simultaneously. Disadvantages of all fixed platforms are that they are more expensive than floating platforms, and they are more susceptible to receiving or causing damage in the event of a collision with a vessel.

Multiple-Buoy Berths

Conventional buoy moorings, also called multiple-buoy berths (MBBs), have been used successfully worldwide in cargo transfer operations with tankers. This type of mooring utilizes several buoys to maintain the vessel in a given position and orientation, as shown in Figure 3-2. Commonly, the tanker's bow anchors are dropped and provide the necessary mooring restraint at the tanker bow. Crude oil is pumped through flexible hoses connected to a submarine pipeline leading ashore.

The main advantage of the MBB is low investment cost. In addition, the conventional buoy mooring is less susceptible to damage by berthing tankers than fixed platforms. However, there is a general tendency to consider the 100,000 dwt tanker the limiting size for this type of berth.

A mooring launch is required with the MBB during both mooring and unmooring; the use of the launch limits operations to six-to-eight-foot seas and winds up to twenty-five knots. Once the vessel is moored, the environmental conditions limiting mooring and cargo transfer operations are similar to those for a sea island.

Single-Point Mooring Systems

Over one hundred single-point mooring systems have been installed throughout the world, making SPM operations established and accepted in the oil industry. SPM facilities are either fixed towers or floating buoys. All SPM systems possess the common advantage that moored tankers are free to weathervane into the prevailing environment, thereby reducing loads. Therefore, ships can remain moored at an SPM in far more severe conditions than at other types of terminals. Vessels can remain moored in seas greater than twenty feet if necessary. The maneuver into and out of an SPM is quicker and safer than with other types of terminals, and does not normally require the use of tugs. In an emergency, a ship can disconnect hoses in under an hour and release mooring lines in fifteen minutes.

While vessels can remain moored at an SPM in severe environments, the berthing operation is limited to six-to-eight-foot seas, due to the usual need to be assisted by a mooring launch. Cargo transfer operations are frequently suspended if seas are more than twelve feet, because of dynamic motion of the hose.

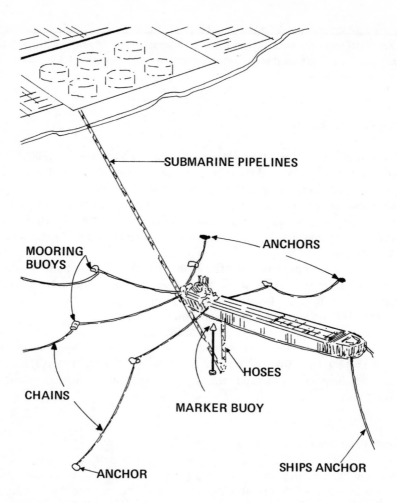

Figure 3-2. Conventional Buoy Mooring. Source: *Volume I, Draft Environmental Impact Statement, Maritime Administration Tanker Construction Program*, N.T.I.S. Report No. EIS 730392 D, 1973, p. IV-205.

Single-point moorings also require more maneuvering room surrounding the berth, which usually results in the berth being further offshore than other mooring alternatives.

Almost all SPMs now in operation are floating buoys called catenary anchor leg systems (CALMs), shown in Figure 3-3. The fixed towers are more expensive than the floating buoys and are more susceptible to damage from berthing

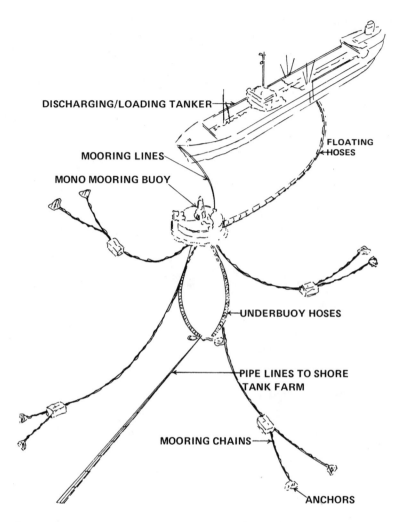

DISCHARGING/LOADING TANKER

FLOATING HOSES

MOORING LINES

MONO MOORING BUOY

UNDERBUOY HOSES

PIPE LINES TO SHORE TANK FARM

MOORING CHAINS

ANCHORS

Figure 3-3. Single Buoy Mooring Facility (CALM). Source: *Volume I, Draft Environmental Impact Statement, Maritime Administration Tanker Construction Program*, N.T.I.S. Report No. EIS 730392 D, 1973, p. IV-207.

tankers. The CALM type of SPM is a cylindrical steel hull ranging in size from thirty-three to fifty feet in diameter, which serves as a stationary floating platform. A turntable or similar device on the buoy supports a rotating mooring bollard to which the vessel is moored. The turntable also supports a rotating pipe arm manifold, to which the tanker-to-buoy floating hose lines are connected.

An undersea hose system connects the under-buoy side of the rotating pipe manifold to the permanent seafloor pipe manifold which, in turn, is connected by submarine pipelines to shore installations. An SPM for supertankers is normally kept in position with six to eight pretensioned chains anchored or attached to piles driven into the sea bottom.

A relatively recent advancement over the CALM is the single anchor leg mooring (SALM), shown in Figure 3-4. The SALM differs in design in that the mooring buoy is anchored to a base on the sea floor by means of an anchor chain which is attached some sixty feet or more below the ocean surface to a

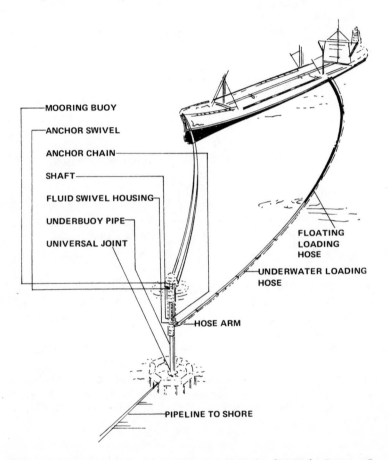

MOORING BUOY

ANCHOR SWIVEL

ANCHOR CHAIN

SHAFT

FLUID SWIVEL HOUSING

UNDERBUOY PIPE

UNIVERSAL JOINT

FLOATING LOADING HOSE

UNDERWATER LOADING HOSE

HOSE ARM

PIPELINE TO SHORE

Figure 3-4. Single Anchor Leg Mooring System (SALM). Source: Joseph D. Porricelli and Virgil F. Keith, "Tankers and the U.S. Energy Situation: An Economic and Environmental Analysis," *Marine Technology*, Vol. 11, No. 4 (October 1974), copyrighted by the Society of Naval Architects and Marine Engineers and included herein by permission of the aforementioned society.

fluid swivel housing at the top of the anchor leg. The cargo hoses are connected to this fluid swivel unit far below the sea surface. Hose wear due to wave action is thus reduced over conventional SPMs, where principal hose wear has occurred at the connection of the hose to the mooring buoy.

The mooring buoy on the SALM is approximately sixteen feet in diameter and twenty-five feet high. Because the SALM mooring buoy can be small relative to the CALM and ruggedly constructed, and because the hose connections are placed below the keel of the tanker, the possibility of serious damage to either the buoy or the vessel as a result of a collision is minimized.

The characteristics of each of the marine terminal mooring systems described are summarized in Table 3-3. Next, the economic aspects of the more popular systems are considered.

Terminal Costs

A recent report by Raytheon Company for the Massachusetts Port Authority estimates the cost of a general SPM system at fifteen million dollars. This cost includes two SPMs, a pumping platform, booster pumps, and site preparation. (The SPM used here is the catenary anchor leg mooring (CALM) type. The single anchor leg mooring (SALM) is expected to be lower in hardware cost, but more difficult to install.) A terminal sited close to shore could eliminate the pumping platform and pumps and their associated cost of five million dollars. Raytheon estimates the cost of a sea island pier as approximately twenty-eight million dollars, which includes site preparation, platform, four breasting and mooring dolphins, walkways, piping, electrical systems, loading arms, and pollution control devices. Exact costs for any specific terminal depend, of course, on the characteristics of the particular site chosen.

Pipeline costs for land and water environment are shown in Table 3-4. The costs of an onshore surge tank farm will vary greatly with size, land acquisition costs, and site characteristics. A range of twenty to fifty million dollars represents most applications; Tables 3-5 and 3-6 show typical annual operating and maintenance costs for SPM and sea island pier terminal systems. Exact costs will depend on weather conditions, amount of traffic, location, and other factors.

Artificial islands vary so greatly in design that it is impossible to generalize accurately about their costs. Probably the most famous artificial island design is that proposed by Soros Associates in their study, *Offshore Terminal System Concepts*, for the Maritime Administration. Known as NADOT (North Atlantic deepwater oil terminal), the terminal was to be located off Delaware. A three-stage construction approach would lead to a total cost of 1.3 billion dollars over an eight-year period.

The initial stage terminal would consist of an island of about 100 acres,

Table 3-3
Marine Terminal Mooring Comparison

Limitations on use	Conventional Pier[c] or Sea Islands	SPM			Multiple Buoy Berth
		Fixed Tower	CALM	SALM	
While berthing					
Waves	3-4	↔	6-8 ft	↔	6-8 ft
Wind	25 kts	↔	25 kts	↔	25 kts
While moored					
Waves	10 from ahead, 3-4[a] ft abeam	↔	Over 15	↔	10 ft ahead, 3 ft abeam
Wind	50 kts	↔	60 kts	↔	20-40 kts[a]
While transferring cargo					
Waves	3-10 ft[b]	↔	10-12 ft	↔	3-10 ft[b]
Wind	30 kts	↔	40 kts	↔	25-35 kts[a]
Distance offshore (Affects pipeline length)	Relatively close	↔	Furthest	↔	Medium
Maneuvering and seabed area requirements	Small	↔	Large	↔	Moderate
Ease in getting under way	Fair/good	↔	Excellent	↔	Poor
Use of tugs	Required	↔	Usually Not Required	↔	Usually not reqd.
Use of launches	Sometimes	↔	Required	↔	Required
Investment	High	High	Moderate	Moderate	Low
Susceptibility to damage	Moderate to high	Moderate to high	Moderate	Low	Low
Ability to handle multiple products	Excellent/good	Fair	Fair	Fair	Poor

[a] Depends on wind velocity and direction.
[b] Depends on wave height and direction. Wave heights are significant.
[c] Convention piers normally protected from waves.

Source: Raymond A. Beazley and Ralph P. Schlenker, "A Rational Approach to Marine Terminal Selection," (October 25, 1973).

Table 3-4
Pipeline Construction Costs

Diameter (inches)	Land Cost/Mile (dollars)	Water Cost/Mile (dollars)
8	$20,000	$200,000
12	45,000	230,000
16	82,000	270,000
20	114,000	320,000
24	148,000	390,000
26	163,000	440,000
28	190,000	490,000
30	200,000	550,000
32	210,000	620,000
36	240,000	770,000
40	272,000	982,000
42	286,000	1,100,000
44	300,000	1,270,000
48	320,000	1,690,000

Source: H.S. Lahman, J.B. Lassiter III, and J.W. Devanney III, *Simulation of Hypothetical Offshore Petroleum Developments,* Massachusetts Institute of Technology, April 1974, p. 25.

Table 3-5
Typical Deepwater Port Annual Operating Costs

Terminal Operating Personnel	$ 224,000
Tank Farm Personnel	320,000
Transportation	135,000
Launch and Assist Vessel Personnel	300,000[a]
Crude Oil Pipelines (land)	120,000
Total	$1,099,000

[a]This cost figure refers to an SPM system. For a sea island pier system, where more vessels are utilized, the cost would be an additional $300,000.

Source: *Massport Marine Deepwater Terminal Study, Interim Report Phase IIA,* Raytheon Company, May 1974, p. 83.

protected from ocean waves by a dog-legged breakwater about 11,500 feet long. The terminal would contain two berths for tankers up to 350,000 tons and would service refineries in the New York-New Jersey area and along the Delaware River. It also would have a pipeline or six shallow-draft berths for 30,000-60,000-ton feeder vessels. At this initial stage, the terminal would cost 499 million dollars and would handle 100 million tons annually.

The next construction stage would double the terminal area, the ship

Table 3-6
Typical Deepwater Port Annual Maintenance Costs

System Component	Maintenance Cost Expressed as a Percent of First Construction Cost
Sea Island Pier	1%
SPM Terminal	3[a]
Tank Farm	2
Land Pipelines	1/2
Submarine Pipelines	1

[a]Does not include costs of inspection and replacement of losses and underwater fittings.
Source: *Massport Marine Deepwater Terminal Study, Interim Report Phase IIA*, Raytheon Company, May 1974, p. 83.

facilities, and the oil-handling capacity, at a cost of 288 million dollars. The final stage, priced at 531 million dollars, would enlarge the island to 500 acres and lengthen the breakwater by 7500 feet. In addition to 300 million tons of oil, the terminal would be able to handle dry-bulk commodities, such as iron ore and coal. It would consist of six deep-draft berths for supertankers, two deep-draft berths for dry-bulk carriers, and thirteen shallow-draft berths for feeder operations (or alternatively a pipeline network).

Because of the high initial costs, artificial islands are not seriously considered when the only cargo to be moved is oil. In those cases where it is necessary to move dry bulk commodities, it would be natural to think of the possibility of an artificial island. However, it takes a special case to make this concept economically (and environmentally) feasible for application in the United States.

Supertanker Costs

The main economic benefit of a deepwater terminal system accrues when it is coupled with supertankers. Economies of scale possible with supertankers are readily apparent in both the construction and operation of the vessels, as was discussed in Chapter 1. The nature of the tanker market from a chartering vantage was discussed, along with the reasons for the acceleration in tankship size and the emergence of the supership.

From the time of initial building of supertankers in the mid-1960s, shippers realized the decreasing costs per unit volume possible with larger tankers. While the trend in larger vessels with lower costs has remained valid, the absolute values have changed. Table 3-7 shows the large increases, as well as the fluctuations, involved in the construction cost of a 210,000-dwt tanker over the years 1966-72. Note that the cost of a supertanker almost doubled between 1969 and 1971.

Table 3-7
Cost of Building a 210,000-DWT Tanker

Date of Ordering	Contract Price	Cost per DWT
End 1966	$13.2 million	$ 63
End 1967	$14.7 million	$ 70
End 1968	$16.6 million	$ 79
End 1969	$19.0 million	$ 90
End 1970	$31.0 million	$148
End 1971	$33.5 million	$159
End 1972	$31.0 million	$148

Source: *The Cost of Ships*, The Research Division, H.P. Drewry (Shipping Consultants) Limited, London, England. 1972.

The great increases in construction costs over the past several years greatly affect the total cost per ton of cargo associated with these tankers. Amortization of capital cost is typically the largest single cost factor for tankers. In addition, some operating costs, such as insurance, are a function of construction costs. Because of this importance of construction cost to total cost, it is difficult to use data from studies before 1971 in analyzing current system costs.

In August 1972 the Institute for Water Resources, within the Corps of Engineers in the Department of the Army, published the *U.S. Deepwater Port Study* by Nathan Associates. Unfortunately, the vessel costs in this report were all calculated to represent costs before the end of 1970. The changes that have taken place in the 1969-71 period in vessel construction costs limit the value of this segment of the study. The in-house report by the Maritime Administration (MARAD), *The Economics of Deepwater Terminals*, and the MARAD-sponsored study, *Offshore Terminal System Concepts*, by Soros Associates, both appear to use more recent cost data than 1970; however, detailed breakdowns of vessel costs are not shown in these reports. Figure 3-5 presents the combined costs of ocean transport, terminal handling, and transshipment.

A study useful for giving the reader an overall understanding of the relationship between vessel costs and size is "Tankers and the U.S. Energy Situation: An Economic and Environmental Analysis," by Porricelli and Keith. As stated earlier, construction cost is an extremely important factor. Figure 3-6 shows the relationship between construction cost and deadweight based on single ship bids from Japan in September 1973 for tankers to be delivered in 1976. These costs are theoretically the same as seen by a United States shipowner purchasing a vessel in a domestic shipyard and benefiting from federal construction differential subsidy funds which make up the difference between domestic and foreign costs. As one would expect, there are savings per deadweight ton in purchasing a larger vessel, since the amount of hull steel, machinery weight, accommodation space for crew, and other factors do not increase proportionally with increases in the size of the vessel.

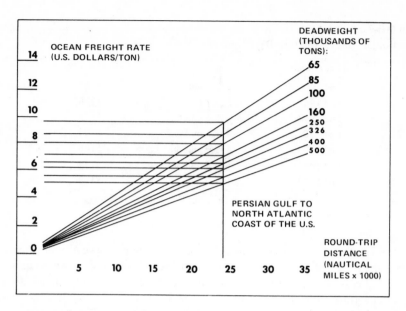

Figure 3-5. Combined Costs of Ocean Transport, Terminal Handling and Transshipment Costs to Shore. Source: *Offshore Terminal System Concepts*, Soros Associates, September 1972.

In a similar manner, almost all operating costs expressed per deadweight decrease significantly with increases in vessel size. The only exception is insurance. Consequently, the overall costs of transporting oil are reduced with larger tankers; this point has been discussed in Chapter 1, and is depicted graphically in Figure 3-7. The required freight rate (RFR) per ton of cargo delivered is the minimum rate which covers all the owner's costs including the return on his invested capital. Figure 3-7, taken from the Porricelli and Keith study, assumes a 10 percent return on investment. This research also uses pre-oil-embargo fuel prices; the impact of quadrupling these fuel rates is discussed later in this chapter.

The operating costs for the Porricelli and Keith study are based on a survey of many different tanker owners, and their data include both U.S. and foreign-flag vessels. It should be noted that U.S.-flag annual crew costs are typically six to eight hundred thousand dollars more than foreign-flag costs. Therefore, while an overall average of both foreign and domestic-flag tankers is useful for an overall understanding of the relationship between vessel cost and size, as shown in Figure 3-7, the resulting costs reflect exactly neither the U.S. nor foreign costs.

The Georges Bank Petroleum Study, produced by the Offshore Oil Task Group at the Massachusetts Institute of Technology, presents 1972 operating cost data for foreign-flag tankers of a variety of sizes, as shown in Table 3-8.

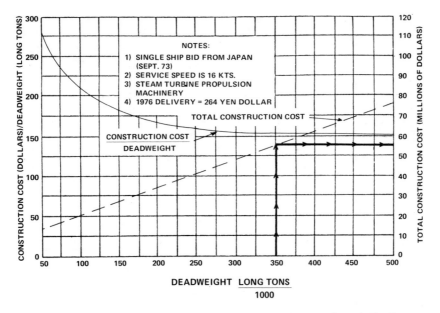

Figure 3-6. Construction Cost vs Deadweight. Source: Joseph D. Porricelli and Virgil F. Keith, "Tankers and the U.S. Energy Situation: An Economic and Environmental Analysis," *Marine Technology*, Vol. 11, No. 4 (October 1974), copyrighted by the Society of Naval Architects and Marine Engineers and included herein by permission of the aforementioned society.

Crew costs vary considerably, depending on the nationality of the personnel. In the MIT study, the "crew costs, which include benefits, repatriation and subsistence, are based on a relatively expensive Western European-Japanese crew. Use of a low wage rate crew (e.g. Spanish officers-Chinese men) could cut crew costs by a factor of two."[3]

In order to understand the relationships between the vessel cost factors, it is helpful to present a percentage breakdown of costs for a specific voyage. A two-part study federally sponsored to provide data for the IMCO meeting shows such information, as given in Table 3-9. While these data are not meant to represent a specific trade route or date, they are useful in viewing the relationship between the various costs. Note that a pre-oil-embargo fuel rate of $23 per ton is used in these calculations.

Table 3-9 shows the relative importance of each cost category. The single most significant cost factor is the annual capital cost or amortization, thus reflecting the capital intensiveness of tanker operations, particularly at larger sizes. Even for the small 75,000-ton tanker, amortization accounts for 46

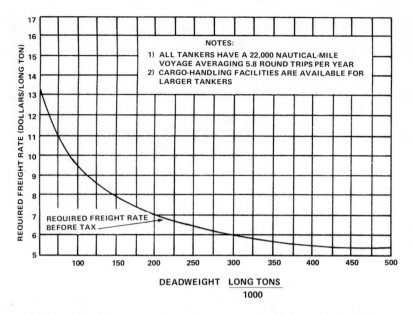

Figure 3-7. Required Freight Rate vs Deadweight. Source: Joseph D. Porricelli and Virgil F. Keith, "Tankers and the U.S. Energy Situation: An Economic and Environmental Analysis," *Marine Technology*, Vol. 11, No. 4 (October 1974), copyrighted by the Society of Naval Architects and Marine Engineers and included herein by permission of the aforementioned society.

percent of total annual cost; for a 500,000-ton tanker, amortization is 58 percent of total cost. Insurance costs become more important as tanker size increases, making up 18 percent of total annual costs for the largest tanker, up from 10 percent for the smallest one. Crew costs become much less important for larger vessels, decreasing from 9 percent for the 75,000-ton tanker to only 3 percent for the 500,000-ton vessel.

Fuel and Tax Variations

Fuel costs, even at pre-oil-embargo rates, also constitute a significant cost factor for tankers which spend the majority of their time at sea. Fuel makes up 24 percent of the total annual costs of the 75,000-ton tanker, decreasing to 12 percent for the 500,000-ton vessel. If the fuel cost per long ton is quadrupled to

Table 3-8
1972 Foreign-Flag Tanker Annual Operating Costs (Thousands of Dollars)

DWT	Crew Costs	Insurance	Maintenance	Administration & Regulation
50,000	350	190	250	200
100,000	360	290	275	200
150,000	370	430	300	200
200,000	380	530	325	200
250,000	390	670	350	200
300,000	400	790	375	200
350,000	410	910	400	200
400,000	420	1,010	425	200
450,000	430	1,090	450	200
500,000	440	1,190	475	200

Source: Offshore Oil Task Group, Massachusetts Institute of Technology, *The Georges Bank Petroleum Study*, Vol. 1, Sea Grant Project Office, Cambridge, Mass., 1973, p. 52.

Table 3-9
Breakdown of Base Ship Economics[a] (Annual Costs as Percent of Total)

Tanker Size (DWT in Thousands)	75	120	250	500
Amortization	46%	49%	56%	58%
Insurance	10%	10%	15%	18%
Fuel cost	24%	21%	15%	12%
Port charges	3%	3%	4%	4%
Manning	9%	8%	5%	3%
Repairs	5%	4%	3%	2%
Prov./stores	4%	3%	2%	2%
Miscellaneous	<1%	<1%	<1%	<1%
Total	100%	100%	100%	100%

[a]Assumes zero tax, preoil-embargo fuel rates, a 10 percent return on investment, and a round-trip voyage distance of 22,000 nautical miles.

Source: Peter M. Kimon, Ronald K. Kiss, and Joseph D. Porricelli, *Segregated Ballast VLCCs, An Economic and Pollution Abatement Analysis*, Society of Naval Architects and Marine Engineers, Chesapeake Section, January 11, 1973, copyrighted by the aforementioned society and used with its permission; and Part 2, *Segregated Ballast Aboard Product Tankers and Smaller Crude Carriers*, Department of Transportation, Coast Guard, February 1973.

approximately reflect post-1973-74 embargo prices, the impact is dramatic, as shown in Tables 3-10 and 3-11. While more efficient supertankers are less affected by fuel cost increases than are the smaller tankers, the quadrupling of fuel cost causes an increase of 36.5 percent in the RFR of the 500,000-ton vessel, and 71.0 percent in the 75,000-ton tanker.

Table 3-10
Impact of Quadrupled Fuel Cost on Required Freight Rate[a]

Tanker Size (DWT in Thousands)	75	120	250	500
RFR with preoil-embargo fuel costs	8.08	6.57	5.36	5.07
RFR with quadrupled fuel cost	13.82	10.78	7.73	6.92
Percent increase	71.0%	64.1%	44.2%	36.5%

[a]Assumes a round-trip voyage distance of 22,000 nautical miles, a 10 percent return on investment, and zero taxes.

Source: Derived from Peter M. Kimon, Ronald K. Kiss, and Joseph D. Porricelli, *Segregated Ballast VLCCs, An Economic and Pollution Abatement Analysis*, Society of Naval Architects and Marine Engineers, Chesapeake Section, January 11, 1973, copyrighted by the aforementioned society and used with its permission; and Part 2, *Segregated Ballast Aboard Product Tankers and Smaller Crude Carriers*, Department of Transportation, Coast Guard, February 1973.

Table 3-11
Breakdown of Base Ship Economics with Quadrupled Fuel Rates[a] (Annual Costs as a Percentage of Total)

Tanker Size (DWT in Thousands)	75	120	250	500
Amortization	27	30	39	42
Insurance	6	6	10	14
Fuel cost	55	52	41	36
Port charges	2	2	3	3
Manning	5	5	3	2
Repairs	3	3	2	2
Prov./stores	2	2	2	1
Miscellaneous	<1	<1	<1	<1
Total	100%	100%	100%	100%

[a]Assumes zero tax, a fuel cost of $92 per ton, a 10 percent return on investment, and a round-trip voyage distance of 22,000 nautical miles. (Percentages may not add to 100 percent, due to rounding.)

Source: Derived from Peter M. Kimon, Ronald K. Kiss, and Joseph D. Porricelli, *Segregated Ballast VLCCs, An Economic and Pollution Abatement Analysis*, Society of Naval Architects and Marine Engineers, Chesapeake Section, January 11, 1973, copyrighted by the aforementioned society and used with its permission; and Part 2, *Segregated Ballast Aboard Product Tankers and Smaller Crude Carriers*, Department of Transportation, Coast Guard, February 1973.

The data presented in the preceding three tables assume no taxes for the foreign-flag operator. (This assumption is not unreasonable for certain foreign-flag operators.) A 50 percent income tax rate has a negative effect on RFR, causing an increase of 24 to 31 percent, depending on the vessel size, as shown in Table 3-12. However, this treatment of taxes can be misleading if applied to the United States tax environment. While a United States vessel owner does pay income tax on his profits, he also can take advantage of various tax shelters. The United States shipowner can benefit from an investment tax credit of 7 percent of the purchase price during the first year, as well as depreciation. The preceding calculations implicitly assume that the shipowner pays the entire purchase price in cash on delivery, while in actual practice, a domestic shipowner typically makes a down payment of less than 30 percent of the purchase price and pays the remainder through means of a mortgage over a twenty- to twenty-five year period. United States tax laws also allow persons to treat interest payments as a tax-deductible expense.[a]

United States flag shipowners may easily conclude that the benefits from tax shelters derived from purchasing a new vessel may outweigh the disadvantages of paying income tax. However, it is important to note that to benefit from the tax shelters created by buying a supertanker, an owner must have enough taxable income, typically in other parts of his business unrelated to this particular supertanker investment, to take advantage of this tax shelter. If an owner does not have such taxable income, the vessel may be leased from a leasing company which has enough other income to fully use such a tax shelter. Under such a leasing arrangement, the leasing company holds title to the vessel and benefits from the related tax shelters. The vessel operator leasing the ship profits also

Table 3-12
Impact of 50 Percent Income Taxes on Required Freight Rate[a]

Tanker Size (DWT in Thousands)	75	120	250	500
RFR with zero tax	8.08	6.57	5.36	5.07
RFR with 50 percent tax	10.07	8.28	6.94	6.62
Percent increase	24.6%	26.0%	29.5%	30.6%

[a]Assumes a round-trip voyage distance of 22,000 nautical miles, a 10 percent return on investment, preoil-embargo fuel rates, and double declining balance depreciation as a tax shelter.
Source: Peter M. Kimon, Ronald K. Kiss, and Joseph D. Porricelli, *Segregated Ballast VLCCs, An Economic and Pollution Abatement Analysis*, Society of Naval Architects and Marine Engineers, Chesapeake Section, January 11, 1973, copyrighted by the aforementioned society and used with its permission; and Part 2, *Segregated Ballast Aboard Product Tankers and Smaller Crude Carriers*, Department of Transportation, Coast Guard, February 1973.

[a]A "net present value" of discounting all cash flows to account for the time value of money is the appropriate methodology to be used to handle all the data. For a discussion of this methodology see Harold Bierman, Jr. and S. Smidt, *The Capital Budgeting Decision* (New York: Macmillan, 1966).

because the leasing company passes on some of the benefits of the tax shelter to the lessee. Through this financial method, a vessel operator can lease a supertanker at a lease rate that is equivalent to a mortgage rate considerably below the prime interest rate. In a sense, the leasing company is paying the vessel operator to undertake the leasing arrangement in order to benefit from the related tax shelters. In conclusion, the effect of doing business in a tax environment may actually be to decrease rather than increase the required freight rate.[b]

Segregated Ballast Designs

The data just discussed are illustrative of basic tanker designs. In conventional operations, such a tanker would travel fully loaded in one direction. On the return trip without cargo, the vessel would place an amount of ballast water in the cargo tanks. When the vessel arrived at the petroleum loading port, the oily ballast water would be pumped overboard.

Ecological concerns have resulted in segregated ballast tanker designs, where the oil cargo and ballast are placed in different tanks. However, increases in environmental safety are arrived at only with a corresponding increase in cost.[c] Studies by the federal government estimate an increase in investment cost of between 4 and 28 percent for various segregated ballast design features.[4] This increase in vessel construction cost requires a higher freight rate to maintain the same return on investment. Under these conditions, the required freight rate typically increases between 3 and 25 percent, depending on the particular situation.

---

[b]For information on present financing arrangements and tax shelters, the reader is referred to "Report to the Commission on American Shipbuilding on a Study of Shipbuilding and Acquisition Financing," by Homer J. Holland and Donald L. Caldera, The First National Bank of Chicago, February 1973, and "Leveraged Leasing: A New Alternative in Financing," First Chicago Leasing Company.

[c]The reader is referred to "Maritime Subsidy Board/Maritime Administration Economic Viability Analysis of the Contract Vessels Pursuant to Stipulation," in *Environmental Defense Fund et al. v. Peter G. Peterson et al.* (U.S. District Court for the District of Columbia, Civil Action No. 2164-72), March 2, 1973. This publication gives detailed analysis of the costs necessary to change the designs of tankers under construction in U.S. shipyards to produce vessels environmentally safer.

4 The Environmental Perspective

Environmental Cost Considerations

From an economic point of view, the need for a deepwater port in the United States appears well substantiated. Yet, despite the apparent economic benefits such a facility would have on the nation and upon the state into whose shores the bulk petroleum would be transferred, no United States deepwater port has progressed beyond the planning stage. State after state has enacted legislation to block possible construction of such facilities. They have determined that it is not in their best self-interest to allow oil companies and other related heavy industry to exploit the potential resources of their coastal zones.

New Jersey is a case in point. Having observed what oil companies had done to the rivers and streams in the northern part of the state, New Jersey residents concluded that it would not be in the state's best interest to let oil companies build a deepwater port facility off the southern coast of their state. They feared the effects of an oil spill, and secondarily, the possible psychological impact on their tourist trade from the presence of an oil terminal.[1]

But even more, citizens of New Jersey feared the detrimental effects that an oil industry would have on their quality of life. While the state of New Jersey is the largest refiner of oil in the United States, contributions to state income from recreational sources are more than double that brought in from oil refining.[2] New Jersey was unwilling to sacrifice the aesthetic value of its shoreline and this important source of state revenue for a smaller source of state revenue and the resultant environmental degradation which it felt would surely follow in the wake of a deepwater port siting.

The state of Delaware suffered a similar plight. In 1969, a consortium of thirteen oil companies, owning over 2000 acres of land along the Delaware coast, made plans to install a major deepwater port in the middle of the bay with oil pipe lines connecting it to existing refineries further up the Delaware River.[3] A similar project was simultaneously proposed by Anglo-Norness, Inc., to build an artificial island in the mouth of the bay for use as a coal transshipment facility. The oil, coal, bulk grain, and steel industries had made extensive plans in this regard, and promised the state of Delaware the greatest growth of industry that had ever been seen in America.[4] Governor Peterson of Delaware placed a moratorium on such construction until the state legislature was able to pass legislation which prohibited this type of activity in the Delaware Bay area.[5]

The state of Maine passed similar legislation in 1970 for the same reasons.

Occidental Petroleum announced plans to build an oil terminal and refinery at Machiasport. Despite Maine's heavy reliance upon this source of energy (90-95 percent of the state's energy comes from petroleum), the Natural Resources Council of Maine and the Sierra Club challenged the accuracy of promised future economic benefits and generated public concern by pointing out the possible effects that oil pollution could have on existing fisheries and on tourism.[6]

A recurring pattern emerges from the study of case after case of proposed deepwater port sitings. First, industry selects a suitable location, draws up tentative plans, and promises great economic benefits nationally, regionally, and locally. Next, regional and local opposition groups emerge and voice environmental concern. Finally, states take legislative action that discourages or prohibits the construction of deepwater port facilities in their coastal waters and/or prohibits the landing of pipelines servicing these facilities on their shores.

In sifting through the vast literature that has accompanied each individual deepwater port siting controversy, the pattern is reinforced through the repeated use of terms that identify the two opposing factions. In general, those who have advocated the construction of the facility have been labeled "economists," while those that have opposed have been termed "environmentalists."[7]

The "economists vs. environmentalists" dichotomy represents the framework most frequently used to study and explain the reasons for differences in opinion at the heart of the controversy. This framework has been used extensively since 1968, when the deepwater terminal issue was first addressed. It is suggested at this point that the "economists vs. environmentalists" framework is far from desirable for objectively assessing and analyzing relevant environmental issues. This point is discussed in great length in the next chapter. The present chapter, however, does not depart from this conventional framework, because the numerous and often voluminous reports, studies, analyses, investigations, proposals, etc. that are condensed and summarized by this chapter are heavily dependent upon it. Because of this heavy dependence, it can be concluded that this chapter summarizes the environmental concerns that were perceived as comprising the major relevant environmental issues during the period from 1968 to 1973.

Offshore Structures

The physical impact of an offshore facility is heavily dependent upon the structure adopted.[8] All offshore structures proposed thus far can be subdivided into one of three categories: artificial islands, fixed structures, and single-point mooring buoys. As in the case of economic desirability, each of the three types has advantages and disadvantages from an environmental perspective.

Environmental concerns with offshore structures concentrate on the initial and long-term effects that the physical presence of each particular structure would have on bottom growth, benthic organisms, aquatic life in the vicinity of the structure, and changes to bottom currents, surface swells, and tidal action. Of the three categories, the artificial island would have the most severe range of effects, while the single-point mooring buoy would have the least.

Construction of an artificial island capable of servicing 250,000-to-300,000-DWT tankers having a length of 1100 feet and a maximum draft of 90 feet would necessitate movement of 94,357,000 cubic feet of sand, gravel, cement and other building materials that would cover approximately 1,130,400 square feet of ocean bottom. Environmentalists have strenuously objected to this type of construction for three reasons. First, the placement of the island itself would immediately destroy approximately 1,130,400 square feet of bottom life, aquifers, benthic organisms, etc. not to mention peripheral destruction around the base of the island structure due to heavy silting and disturbed sedimentation. Second, the back-and-forth traffic required to transport and deposit the building materials would disrupt normal ecological activity in the water column and on the ocean floor beneath the traveled routes. Third, given that conventional practices would be used in obtaining the required building materials, quarrying activity ashore and/or dredging activity in inland and coastal waters would result in massive environmental destruction and degradation.

While these arguments may seem extreme to some, they are based on the fact that man is extremely dependent upon his environment and knows little about the ecological interrelationships of marine life and their ultimate impact on his own. It is to this basic premise that all environmental concerns can be traced.[9] Taking their departure from this point, the environmentalists have been quick to point out that man knows least about the ecological activities in his coastal waters. Construction of such a facility could result in the destruction of valuable feeding and breeding grounds with resultant detrimental effects to various sea life, fish, and shellfish populations. Further, the presence of such a structure would alter existing ocean currents, swell patterns, and surface movements, and affect tidal action, which could produce harmful effects on valuable and ecologically fragile marshlands and estuaries considerable distances away from the island structure. Accurate assessments of current movements, etc., can only be made if information is obtained on "the tidal flow and tidal prism, the water depth, size of water body and water inflow."[10]

On the other hand, the island itself could be designed, shaped, and constructed to minimize its effects on water movement. Logically, it should be possible to design and construct an island that would actually be environmentally beneficial. This type of an approach, however, raises the question as to what is and is not considered environmentally beneficial.[11] Most reports avoid those areas which require such qualitative, highly subjective determinations. They frequently touch upon the objective argument, however, that the presence of such an island would provide additional feeding and breeding areas for underwater life. Environmentalists have also studied this possibility and concluded that while forms of aquatic life might congregate in the vicinity of an artificial island (making it a sports fisherman's paradise) its presence would have little effect on the overall fish population (with or without sports fishermen).

The environmental impacts produced during construction of an offshore fixed structure are the same in nature but somewhat less severe than those that would be incurred during construction of an artificial island. The degree of severity is determined by the amount of underwater blasting that would be required to prepare the ocean bottom to receive the supporting members of the structure. Once in position, however, the structure would have very little effect on bottom currents, silting, etc. The only long-term effect might possibly involve turbulence introduced in the immediate vicinity of the underwater supporting members.

The single-point mooring buoy would have the least environmental impact of the three. The only disruption would occur during placement of the monobuoy's anchoring devices on the ocean bottom. Long-term impacts would be virtually nonexistent; with the exception of spills.

Transfer Methods

One last consideration associated with all three types of offshore terminals is the impact of transferring the bulk petroleum from the terminal to a storage or refining facility ashore. Transfer methods can be divided into two categories: a barge shuttle operation or a submerged pipeline operation. The submerged pipeline method can be further subdivided into above-ground or buried pipelines.

The barge transfer system is highly unsatisfactory from an environmental perspective. The multiple handling required by this method greatly increases the likelihood of spills and other accidents, due to the increased number of couplings and uncouplings. The potential for collisions, rammings, and groundings is greatly enhanced also. Further, this method requires the acquisition of additional property at the shoreline to facilitate the transfer of the bulk petroleum from the barges to land-based pumping stations, storage tanks, or refineries.

Pipelines are considerably more desirable. Of the two categories of submerged pipelines, the buried type is probably the less environmentally disruptive. Pipelines on the seafloor could affect bottom currents, cause siltation and turbulence, and, of course, are constantly threatened by ship's anchors. Both types of pipelines are exposed to hazards such as substrata shifting (i.e. earthquakes, turbidity currents, etc.) which could damage and/or rupture them, with resultant environmental damage. However, the buried, submerged pipeline should have little effect on its environment except for temporary disruption and some destruction of marine benthic organisms when they are initially laid.[1 2]

Onshore pipelines, even though buried, will still have a disruptive effect on residential and commercial areas during their construction. These areas would also be exposed to additional potentials for environmental degradation (i.e., contamination of groundwater, etc.) should any leaks develop that are hard to detect and difficult to locate.

Shore Based Facilities

Ultimately every pipeline stops somewhere. In the case of a deepwater terminal pipeline the end must constitute a storage site or refinery. Both are environmentally undesirable.

The rupture or collapse of storage tanks can result in major pollution incidents. In 1970, the United States Coast Guard recorded eight such incidents involving approximately 3.5 million gallons.[13] Pollution of this type occurs when the petroleum enters the surrounding ground environment, thus affecting algae, bacteria, and other microscopic species.

Refineries are environmentally disruptive on a continuing basis. They create air pollution through the emission of hydrocarbons, sulphur compounds, nitrogen oxides, and carbon monoxide. In addition they pour out undesirable odors which are aesthetically displeasing. Refinery operations also require large volumes of water. It is estimated that seven gallons of cooling water are required to process one gallon of gasoline.[14] The water used in the actual processing operation is thermally polluted and may become "contaminated with oil and many highly toxic chemicals."[15] The polluted effluent may be discharged partially treated or untreated into a convenient water stream, resulting in more environmental degradation.

The problems associated with storage tanks and refinery operations are not insurmountable. Stringent air quality standards might ensure that petroleum refineries install and correctly utilize the equipment necessary to reduce air pollutants. The need for large quantities of cooling water and the associated problems of partially treated and thermally contaminated effluents might be ameliorated through recycling used waters. Both storage tanks and refineries could be designed and located in harmony with the needs of other coastal zone users. The "sensitivity of resident biota to oil contamination"[16] and the multifaceted environmental effects of secondary and tertiary build-up could be considered during terminal design and development. Storage facilities can blend into a landscape, be buried or generally hidden from view.[17] It can be concluded, therefore, that the problems presented thus far within the confines of the environmental perspective, while difficult, are solvable.

Vessel and Terminal Operations

Based on information presented in most, if not all, deepwater port literature published to date, it appears obvious that the vessel and terminal-operation facet lies at the very center of the deepwater port controversy. Environmentalists vehemently object to the environmental degradation that would result from the spillage of oil. Their concerns can be subdivided into two categories: those associated with one-time catastrophic releases of oil into the environment, and those associated with recurring minor releases that result from normal vessel and terminal operations.

The catastrophic type of incident is probably best typified by the *Torrey Canyon* disaster. On March 18, 1967, the 119,000-DWT tanker *Torrey Canyon* struck a reef near Cornwall, and subsequently released 117,000 tons of crude oil.[18] This incident resulted in damages in excess of $10 million in as-yet-unresolved litigation, not to mention irreparable ecological and environmental damage.[19] Another incident frequently cited is the grounding of the *Ocean Eagle.* The rupture of this small 12,065-DWT tanker caused damages in excess of $2 million to valuable resort beach area in Puerto Rico.[20]

Congressman Roth of Delaware summarized the feelings stirred by this type of concern in the following way:

If only one of these supertankers runs aground, if one of these supertankers hits a small barge with a resulting leak, or if a flash fire breaks out on board only one of these vessels—as has happened off the Coast of Africa—the damage to Delaware's coastline could be irreparable.... I shall also point out that an oil catastrophe could not only damage our shoreline and environment, but endanger the livelihood of thousands of our citizens.[21]

Fears of this type are justified. The tankers that would be serviced at the deepwater terminals that have been proposed would be considerably larger than the *Torrey Canyon.* Vessels such as the *Idemitsu Maru, Nisseki Maru,* and the *Globtik Tokyo,* whose massive dimensions are arrayed in Figure 1-7 (p. 22), are representative examples of tankers that could be serviced by a deepwater terminal.[22]

The potential for casualties involving these large crude carriers is also very real, especially during rough weather. Large tankers are difficult to maneuver and have longer crash-stop distances (which may even exceed two miles).[23] These factors can be considered significant, in view of the fact that casualties and accidents involving much smaller conventional vessels already account for an estimated 11 percent of the entire ocean oil pollution.[24]

In any event, an oil spill from a large crude carrier along the coastline of the United States would in all probability have catastrophic results. As stated in the March 1968 Report to the President, submitted by the Interior and Transportation departments: "This country is not fully prepared to deal effectively with spills of oil or other hazardous materials—large or small—and much less with a *Torrey Canyon* type disaster."[25]

The second type of operational pollution resulting in the course of normal vessel and terminal operations, while not as dramatic, is much more significant.

It is estimated that the discharge of oil-contaminated ballast and wasting residues, along with bunkering and minor leaks, accounts for 34 percent of all oil pollution in the oceans.[26]

As environmentally undesirable as these almost certain consequences of deepwater terminal operations may seem, they are to a large degree correctable or preventable. Normal operating spills can be reduced to a significant extent through effective design of the facilities and vessels, establishment of terminal operating procedures, adequate training for operating personnel, and vigorous, well-exercised inspection programs.

The establishment of traffic control systems to monitor vessel movements in and out of terminal areas and the installation of electronic communication equipment and special navigation devices on board deep-draft vessels should also reduce the danger of collisions between ships and reduce the likelihood of rammings and groundings in the vicinity of the terminal.

Efforts are currently under way to find ways to alleviate the potentially harmful effects of oil spillage, by preventing such spillage and by the speedy recovery and/or clean-up of spilled oil. Attempts to alleviate oil discharges encountered during the course of normal tanker operations include the introduction of the "load on top" technique and development of clean ballast systems.

This technique involves the loading of new cargo on the residue of oil that was previously partially separated in a slop tank for contaminated water. Unfortunately the success of this method to date is limited, because subjective visual observations are used to determine what proportion of oil to water is clean enough to pump over the side and because only 25 percent of petroleum companies employ this technique due to the reluctance of some refineries to accept the salt-water-contaminated cargo.[27]

Ongoing research has generated several alternatives to alleviate the problems associated with segregated and unsegregated ballast. Some of these include improvements in the construction and design of shipboard tanks, development of recorders, interface detectors, and automatic shutdown devices, and development of physical methods of ballast and cargo separation. With respect to the last method, a joint United States Coast Guard and Navy study is currently testing the feasibility of using rubber membraneous linings in cargo tanks for segregating oil and ballast water. Other alternatives include the provision of slop treatment equipment at the terminal facility, automatic leak detection and shutdown devices, and oil containment barriers.

On the clean-up side rapid response methods have been designed for the purpose of containing and removing oil once a spill has occurred. Containment is necessary to prevent spreading over an extensive area and increasing the potential for environmental damage and the difficulty of clean-up. Containment apparatus include pneumatic barriers and mechanical booms. Pneumatic barriers operate by forcing compressed air through perforated tubes, setting up counter-currents at the water surface to prevent the spread of oil. Mechanical floating booms are curtain barriers of one or two feet in depth. Containment systems have limited use, however, especially in strong currents or waves. The Coast Guard has developed several systems for oil pollution control on the high seas. A typical oil-containment boom was designed for 4-foot seas, 2-knot currents, and 20-mph winds.[28]

The actual removal of oil can be accomplished through physical or cosmetic means:

The removal of an oil spill upon water may be accomplished through the actual physical removal of the oil—the preferred method from an environmental viewpoint—or through the cosmetic removal of the oil from the water's surface. Complete removal can be accomplished by skimming devices, burning or the application of absorbents. Cosmetic removal is accomplished via dispersants or sinking agents.[29]

The traditional form of oil-spill treatment has been to use detergents, not to destroy the oil but to disperse it into small oil droplets, thus facilitating natural degradation. The unsatisfactory nature of dispersants as an oil spill amelioration method was demonstrated in the *Torrey Canyon* incident, when the Plymouth Laboratory of the Marine Biological Association of the United Kingdom concluded that the chemical dispersants used to clean up the oil were many times more toxic than the oil itself and therefore more environmentally harmful. The liberal application of over 10,000 tons of detergents was destructive to living organisms.[30]

Physical removal of oil, on the other hand, is accomplished by skimming devices—floating vehicles which separate surface oil from the water mixture. Skimming is usually considered the best method. Burning is not favored because it creates air pollution and is difficult to execute because of the complex problem of maintaining the temperature of combustion in cold water.

Absorbents can be an acceptable method of physical removal. Of the many absorbents available, straw is one of the best.[31] Problems still arise, however, as to what to do with the expended straw since burning it results in air pollution and dumping produces contamination of the water or land fill into which it is dumped.

The problems of oil clean-up are compounded if the spill plume reaches shore. Measures to clean polluted beaches also include the use of absorbents, such as straw, which are subsequently removed by bulldozers along with the surface layers of contaminated sand. Once again the problem of disposal is raised, since disposition of contaminated absorbents and oily sand contaminate everything they contact, ultimately allowing the oil to resurface.

Oil Spills

During the period from 1968 to 1972 there was considerable question as to how environmentally significant an oil spill actually was. While the immediate impacts of the *Torrey Canyon* spill were obvious and dramatic, the intermediate and long-term effects seemed to be nonexistent. Straughan conducted a twelve-month study on the oil well blowout that occurred in 1969 in Santa Barbara Channel. Despite the amount of oil that was lost to the environment, he detected neither a change in the ecological balance, nor a measurable effect from oil pollution.[32]

Studies dealing with the long-term environmental impacts of small recurring oil spills, like those associated with normal vessel and terminal operations, show similar results. Crapp studied the "biological effects of terminal operations at Bantry Bay," and concluded that there was nothing to report because no oil pollution had been detected.[33]

Most laboratory studies, however, tend to refute these findings. As previously stated, the immediate environmental impacts are readily apparent. The clinging characteristic of oil caused the death of thousands of birds in the *Torrey Canyon* spill and the Santa Barbara blowout.[34] The oil reduces "the buoyancy of the birds and interferes with the protective layer of trapped air in the plumage, thus leading to death from drowning or exposure."[35] In addition, "oil deposited on the coast is severely damaging to coastal amenities and may affect shore animals by smothering them."[36] While not as dramatic, and hence less evident, laboratory experiments suggest that the intermediate and long-term effects of oil pollution can be even more catastrophic.

The difficulty in assessing the true impacts of any oil spill results from the vast number and complexity of interactions between largely uncontrollable variables. The eventual environmental impact due to a spill is a function of a great many variables that are far more subtle than type of oil, amount, geographic location, and weather conditions. The toxicity of crude petroleum, for example, depends on its source where variations exist because of given physical/chemical properties, e.g. volatility, specific gravity, and boiling point. These variations are manifest in the contrasting impact of the widespread destructive effects on marine life of an oil spill off West Falmouth, Massachusetts[37] as compared to the minimal effects and rapid regenerative capability of marine organisms in the Santa Barbara oil blowout.[38]

Continuing with this example, the long- and intermediate-range impacts of the toxity of oil on marine life are not immediately apparent. This is due to the sometimes extremely subtle nature of the environmental disruptions caused by an oil spill. Most marine organisms depend on complex behavioral signals to maintain normal life patterns such as feeding and reproduction. Extensive studies of marine life and its dependence on these chemical signals (called *pheromones*) have focused on migration habits and territory recognition by fish;[39] feeding, reproduction, and social behavior in fish and lobsters;[40] predation by starfish.[41] It has been discovered that the utilization of pheromones in extremely low concentrations, measured in parts per billion, is recognized by marine organisms.[42] This raises some serious questions as to the effects toxic materials might cause by blocking, masking, or substituting false chemical signals.[43] Environmental disruptions in these areas, feeding, migration, and reproduction, could endanger the survival of large populations of marine life, causing further disruptions in food chains.

Incorporation of hydrocarbons in the food chain comprises another set of parameters that merits serious consideration. Crude oil contains toxic compounds that affect both marine organisms and human beings. Blumer has demonstrated that "hydrocarbons derived from petroleum do get into marine organisms and are passed on in the marine food chain until they reach us in food."[44] This transport of hydrocarbons in the marine food chain points up the fact that the sea's ability for recycling wastes is finite. Indeed, the presence of toxic compounds of the magnitude occurring in frequent large oil spills places intolerable strains on marine ecosystems.

Of further consideration is the fact that hydrocarbons contained in petroleum are believed capable of producing anesthesia and cell damage.[45] It is also believed that they are carcinogenic. Blumer, in a public hearing before the Special Commission on Marine Boundaries and Resources, Commonwealth of Massachusetts, cited the report of the British Medical Research Council in 1966:

It has been demonstrated by means of biological tests that many of the oil fractions (constitute) evidence that there are carcinogens in the crude oil for throughout the investigation every precaution has been taken to avoid thermal decomposition and chemical rearrangement of the compounds present in the crude oils. Chemical identification of many compounds has been achieved and a few of these are capable of producing tumors.[46]

Once again, the claim by environmentalists that man knows too little about the ecological interrelationships of marine life and their ultimate impact on human beings seems justified. Because of this they further reason that change should not be introduced arbitrarily.

The Environmental Perspective

While the "economists vs. environmentalists" dichotomy has provided a convenient framework from which to discuss the ongoing controversy, it has done little toward providing any understanding as to whether or not a deepwater terminal is necessary and/or desirable, economically or environmentally. The major shortcoming of this frequently used framework is that it attempts to compare *quantitative* data based upon *speculative* projections against *qualitative* information based upon *subjective* judgments. Thus, all the key issues on both sides are addressed but never compared. The framework provides no mechanism for weighing projected economic benefits, measured in dollars, against environmental concerns, measured in degrees of fear and uncertainty.

In summary, this chapter has presented the most important concerns that relate to the deepwater port siting controversy from an environmental perspective. During the period 1968 to 1973 these concerns centered on such topics as: (1) choice of structure utilized as that choice relates to aesthetics, probability of spills, and disruptive effects on the tidal prism; (2) choice of transfer methods as they relate to probability of leaks or spills, susceptibility of pipelines to damage, and aesthetics; (3) vessel and terminal operations as they relate to probability of collisions and resultant spills; (4) the long-term effects of one-time catastrophic spills and recurring minor spills as they in turn effect aquatic life, shoreline aesthetics, and the well-being of those who live in the coastal zone area; (5) oil spills as they relate to possible disruptions to acquatic feeding, spawning, and social behavior, with resultant disruptions to food chains; and (6) oil spills as they relate to possible alterations in the food chain which could be harmful to human life. For these reasons environmentalists have concluded that no deepwater terminal should be constructed in United States waters until the direct and ancillary effects brought about by deepwater terminal construction can be established and projected with greater certainty.

Issues, Actors, and Onshore Impacts

Understanding the Problem

The "economists vs. environmentalists" dichotomy does have serious short-comings. First, it is entirely too simplistic (which is probably the very feature that has led to its quick adoption and usage). Second, it lends itself too easily to the formulation of broad generalizations which lead, in turn, to false conclusions, incorrect evaluations and assessments, and the formulation of incomplete and sometimes improper solution to problems that are often more apparent than real. Last, and most important, it tends to exaggerate environmental considerations far beyond their bounds of reasonableness. This is not to say that environmental considerations are relatively unimportant or insignificant factors in the deepwater port siting controversy. Obviously, quite the contrary is true. Environmental considerations by themselves, however, would probably have not delayed the site selection and construction of a deepwater port thus far. ("Environmental" in this context refers to the primary physical impact of building and operating a deepwater terminal, but does not include the secondary impacts onshore.)

The following quote serves to emphasize the point:

Although there is considerable support for handling large tankers offshore in preference to congested North Atlantic ship channels, increased public and political awareness of oil's potential as a pollutant and its damaging effects on local ecology, natural resources and valuable recreational areas, is presently the most significant constraint to deepwater port development. . . .
Furthermore, the idea of deep draft bulk carriers in these harbors tends to constitute a threat that transcends the potential physical damage they can cause. Although this fear is largely psychological in origin, it nevertheless constitutes a significant factor that would have to be dealt with in any plan for deep draft bulk carriers entering North Atlantic ports.[1]

If the threat of oil pollution does indeed constitute the only obstacle of any major consequence, then most certainly this "threat" would transcend any possible physical damage that any in-depth economic analysis would show. The point being made here, however, is that misplaced emphasis on environmental concern has lead to the erroneous conclusion that "fear of oil pollution" is "the most significant constraint to offshore port development in the North Atlantic for large bulk carriers."[2]

This contention is heavily reinforced by a statement by Governor Russell W.

Peterson of Delaware: "As far as I'm concerned, even if Shell Oil can build a plant 100% free of pollution, I'm still opposed."[3] Peterson goes on to say that Delaware has elected to preserve its shores for tourists, recreational, and compatible industrial uses.[4]

Congressman Sandman of New Jersey expressed this line of reasoning much more explicitly when he stated:

Establishing an oil terminal in lower Delaware Bay, in my opinion, whether there is spillage or there is not, is still objectionable and strenuously opposed by the half million people I represent from my district.

They may have little concern whether there is going to be spillage of oil. That is not their major concern, even though it is a great concern. They are concerned with what is going to be brought in with this particular facility. It is only going to be a foot in the door. First an oil transmission line, then later a big marine terminal for oil tankers, and then sometime after that there will be more places to have some oil refineries there. And I am confident that this part of the country does not want any oil refineries in that particular area.[5]

Quite obviously, though "fear" might be a significant motivating factor in rallying opposition to deepwater terminal development, the role played by "fear of oil pollution" may be small. Similarly, while concern for local and regional ecosystems and a sincere concern for the natural environment constitute very real issues, their roles in stirring opposition to deepwater terminal development could conceivably be minor. It is the purpose of this chapter to identify issues, actors, and the roles they play in locating deepwater terminals. This chapter concludes with the development of a more functional and useful framework for conceptualizing, understanding, analyzing, and evaluating these central issues.

In summary, while environmental concerns constitute one very important issue in the deepwater port siting controversy, they should not be allowed to overshadow other issues whose impact might be much more significant. In addition to the environmental question, other important aspects to be considered include legal, political, and socioeconomic issues.

Legal Issues

Legal issues play a very significant role in determining how, why, what, when, and where a deepwater terminal will or will not be permitted. They run the gamut from minor to major impacts in all areas of concern in the siting controversy. Many questions of law remain unresolved at all levels: international, national, state, and local. These questions can only be resolved as deepwater terminal proposals pass through the intricate legal maze. This process is a lengthy, time-consuming one, in and of itself. This chapter only touches upon a representative sampling of the legal questions that are raised in an effort to convey some feel for the complexity of legal issues involved. The following chapter deals with pending legislation.

The legal issues will be considered with regard to three categories: siting, liability, and ownership and control. In examining first the siting of a deepwater terminal, some of man's basic concepts of ocean surface appropriation, ocean resource exploitation, jurisdiction, and sovereignty are brought into question. Grotius is credited with advancing the idea that the oceans be considered as free, the common property of all men, *res communis*.[6] This runs in stark opposition to the concept used by nations in acquiring unappropriated land, *res nullins*.

Exceptions to the *res communis* rule have always been accepted. The three-mile territorial sea, originally determined by the distance a cannon ball could be fired, has been used as the norm by most nations for several centuries. Thus, appropriation of territorial sea area for deepwater port purposes is clearly within limits of existing international law, and would be subject to regulations by municipal legislation.[7]

In the recent past some nations established contiguous zones of an additional distance, usually nine miles, beyond their territorial sea, within which immigration and custom laws could be enforced.[8] The establishment of these zones helped set the stage for the more dramatic legislation that has followed. As increasing technology enabled certain nations to exploit the natural resources of the sea beds of the Continental Shelf, demand for legal justification of this exploitation increased. In 1945, the United States took unilateral action and created the legal justification through the Truman Proclamation.[9] This proclamation established the inherent right of the United States as a coastal nation to freely exploit the resources of its adjacent continental shelf.

The Truman Proclamation . . . came to be regarded as the starting point of the positive law on the subject, and the chief doctrine it enunciated, namely that of the coastal state as having an original, natural and exclusive (in short a vested) right to the continental shelf off its shores, came to prevail over all others.[10]

In 1958 the central elements of this proclamation were incorporated in articles 1 and 2 of the Geneva Convention on the Continental Shelf.[11]

Having established the precedent of unilaterally exercising sovereign power for the purpose of utilizing the resources of the coastal zone, other nations quickly followed the lead of the United States. Many nations, most notably the South American countries, established territorial waters up to 200 miles beyond their shores.

Thus far, international legal concern has focused on exploitation of natural resources, and has only anticipated the interference that results from such exploitative activities.

Unfortunately the erosion of the freedom of the seas does not stop here. We have already referred to the obvious danger of increasing interference with the surface of the high seas by the expanding exploitation of seabed and subsoil resources. It is almost inevitable that sovereignty over the resources of the seabed and the subsoil should extend upward, that the coastal states should regard it as their prerogative and duty to protect the installations and structures erected under their jurisdiction and that the rights of other nations in these zones should become licensed.[12]

Thus, it can be concluded that international law of the sea could play a definitive role in the deepwater terminal controversy. This overly simplistic discussion of the law serves to point out several possibilities: (1) the terminal must be sited within the limits of the territorial sea; (2) the limits of the territorial sea might need to be expanded to encompass planned deepwater port sites; and (3) political factors would have to be assessed against potential gains and losses in the international political arena if the port were to be sited outside territorial waters.

Turning attention to national and state legislation, the Coastal Zone Management Act of 1972 is a significant law that has already affected the siting of deepwater terminals.[13] This act essentially encourages states to develop their own coastal-zone management program.

The act defines the "coastal zone" as:

The coastal waters (including the lands therein and thereunder) and the adjacent shorelands (including the waters therein and thereunder), strongly influenced by each other and in proximity to the shorelines of the several coastal states, and includes transitional and intertidal areas, salt marshes, wetlands and beaches. The zone extends, in Great Lakes waters, to the international boundary between the United States and Canada and, in other areas, seaward to the outer limit of the United States territorial sea. The zone extends inland from the shorelines only to the extent necessary to control shorelands, the uses of which have a direct and significant impact on the coastal waters.[14]

Continuing, Congress defined "management program" in the following way:

"Management Program" includes, but is not limited to, a comprehensive statement in words, maps, illustrations, or other media of communication, prepared and adopted by the state in accordance with the provisions of this title, setting forth objectives, policies, and standards to guide public and private uses of lands and waters in the coastal zone.[15]

This act has focused the states' attention on the fact that deepwater terminals comprise but one major interest in competition with others for the scarce resources of the coastal zone. Many states recognized this fact several years earlier and initiated programs to coordinate the planned development and usage of their coastal zones. The Coastal Zone Management Act, therefore, merely provided added incentive in the form of financial assistance to these states and others to develop comprehensive, formal coastal-zone management programs. The existence of such a program meant that the desirability of having a deepwater terminal sited in any state's coastal zone would be given a hard look with respect to the potential advantages and disadvantages the terminal and its associated secondary build-up would bring in comparison to those of other competing uses.

Examples of state laws predating the passage of the Coastal Zone Management Act include the Massachusetts Coastal Wetlands Protection Act, the Inland
Wetlands Dredge and Fill Law of 1965, and the Inland Wetland Protection Act
of 1968 and Maine's Coastal Wetland Regulation Act and Coastal Conveyance of
Petroleum Act. These pieces of legislation were aimed at preserving state
wetlands which were regarded as their single most valuable coastal resource in
view of the relationship of such areas to the propagation of aquatic life such as
shrimp, clams, lobsters, waterfowl, fish, etc.[16] For this reason similar legislation
was passed in the states of Maryland, Oregon, Wisconsin, California, Hawaii,
Delaware, Washington, Rhode Island, New Hampshire, Connecticut, New York,
New Jersey, Virginia, North Carolina, Georgia, Flordia, Mississippi, Louisiana,
Texas, Minnesota, and Alaska.[17]

But wetlands preservation was not the only motivation. In California, Texas,
Oregon, and Hawaii, public-beach rights clashed with private interests in the
ownership of shore-front property. Legislation passed in these states guaranteed
the public's access to beach areas that have a long-standing history of use for
public purposes.[18]

Both wetlands preservation laws and public-beach access laws have had an
impact on deepwater port site selection. Both severely limit the number of
available siting options. Further, many states have passed legislation that deals
directly with either power-plant or deepwater port siting. States in this category
include Maine, Rhode Island, Delaware, Maryland, Washington, Oregon, California, and New York.[19] Legislation of this type has had a further limiting effect
on potential deepwater terminal development. This concludes the discussion of
siting and the law, the first general area of legal issues.

Liability is the second general legal area to be covered. The discussion now
will focus on existing laws and legislative remedies that enable individuals,
corporations, and/or certain class-action groups to receive compensation for
injuries resulting from maritime torts. It serves to point out the fact that, under
the present system, determination of liability and identification of compensa-
tory mechanisms for injuries resulting from torts which arise during the course
of deepwater terminal operations would most assuredly require a very complex,
tedious, involved process. This suggests that definitive legislation outlining legal
and administrative remedies, and setting the scope and depth of the liability of
those concerned (i.e. owners, vessels, terminals, insurers, etc.) might be a
prerequisite for deepwater terminal development.

Under the present system, offshore deepwater terminals fall under the
admiralty jurisdiction of the federal government. Since all cases falling under
admiralty and maritime jurisdiction (which extends to all navigable waters,
whether interstate or international) are subject to the admiralty power[20] of the
federal courts,[21] all torts that occur at offshore deepwater terminals are subject
to admiralty jurisdiction. It is usually accepted that the place where the injury
occurs and not the place where the tort was committed determines the
jurisdiction of a court over a case. The Extension of Admiralty Jurisdiction

Act[22] extends admiralty jurisdiction to all cases of injury to persons or property caused by a vessel in navigable waters, even though the injury may be suffered on land.

Parties can proceed *in personnam* against a natural or corporate person or *in rem* against a ship. A suit *in personnam* may be subject to the Limitation of Vessel Owner's Liability Act.[23] This act provides that a vessel owner's liability for loss, damage, or injury by collision or for any act done, occasioned, or incurred without the privity or knowledge of such owner, shall not exceed the amount or value of the interest or knowledge of such owner in such vessel, and her freight then pending.[24] Had a suit been levied against the owners of the *Torrey Canyon*, for example, (whose grounding and ultimate destruction resulted in over 16 million dollars in damages) under this theory, their liability would have been limited to fifty dollars; the value of one lifeboat that was salvaged.

Suits *in rem* are usually employed against vessels having foreign ownership and registry. These suits may be brought only for the purpose of enforcing maritime liens.[25] Liens arise either by violation of statute, like the Federal Water Pollution Control Act,[26] or from the violation of general maritime law.

At this point it should be noted that existing legislation is wholly "vessel" oriented. It is not certain, therefore, just how a deepwater port facility would be treated under the law. Obviously the type of structure adopted will influence the interpretation of the term "vessel." That is, a single-point mooring buoy would probably meet the definition of a vessel and existing law could easily be adapted, while an artificial island would not, and existing law could therefore not be applied.

In turning to questions of liability and its impact on deepwater terminal development, it becomes apparent that both existing laws of long standing and recently passed legislation have played major roles in setting the pace of deepwater terminal development in this country. In the past, long-standing laws that have significantly affected the development of deepwater ports were: the Rivers and Harbors Act of 1899[27] and the Federal Water Pollution Control acts of 1948, 1956, 1961, 1965, and 1966 and amendments. More recently passed legislation which deals more directly with issues raised by deepwater terminal proposals includes: The National Environmental Policy Act of 1969, the 1970 and 1972 amendments to the Federal Water Pollution Control Act, the Freedom of Information Act, the Noise Control Act of 1972, The Clean Air amendments of 1970, and the Coastal Zone Management Act of 1972.[28]

From the standpoint of liability, the oldest act having a significant impact in the past would be the River and Harbors Act of 1899, the most recent would be the 1972 amendment to the Federal Water Pollution Control Act. An understanding of both is helpful in developing an understanding for what present liabilities are and who might most likely incur them as a result of torts arising from deepwater terminal operations.

The Rivers and Harbors Act, more popularly referred to as the Refuse Act,[29] makes it unlawful to discharge or deposit refuse matter in any way, or cause refuse matter to be discharged or deposited,

from or out of any ship, barge, or other floating craft of any kind, or from the shore, wharf, manufacturing establishment, or mill of any kind or description whatever other than that flowing from streets and sewers and passing therefrom in a liquid state, into any navigable water of the United States.[30]

Guidelines issued by the Justice Department limited prosecution under the act to significant discharges that were accidental or infrequent. Discharges of a continuous nature were handled under the Federal Water Pollution Control Act of 1970.[31]

A definition of refuse matter was provided by Justice William O. Douglas in *United States v. Standard Oil Co.*[32] Standard had argued that their accidental discharge of commercially valuable aviation gasoline into a navigable waterway did not constitute a violation of the Refuse Act. Aviation gasoline was, after all, a valuable commodity and therefore could not be considered refuse matter. Justice Douglas noted that the aviation gas, once spilled, was not of much value to anyone. Most important, he felt that a "narrow, cramped reading" of the act would defeat its purpose.[33]

An interpretation of the meaning of the street-and-sewer run-off exception came forth in *United States v. Republic Steel Corp.*[34] An accidental release of oil had entered street storm sewers and eventually reached a navigable waterway. Republic Steel argued that since the oil had entered the navigable waterway via the sewer system, the exception applied. The court held that refuse flowing from street and sewers meant sewage, essentially organic waste matter that eventually decomposes. Oil could not be considered in this category.

Violations of the Refuse Act are punishable as misdemeanors. Violators are subject to a fine of not more than $2500 or less than $500, imprisonment for not less than thirty days or more than one year, or both.[35]

The Federal Water Pollution Control Act of 1972 as amended essentially replaced the Oil Pollution Act of 1946. The 1972 act defines an offshore facility as any facility other than a vessel, located in, on, or under any of the navigable waters of the United States. The authority of this act extends to vessels outside the territorial waters of the United States and to vessels and facilities inside the territorial waters. The act prohibits the discharge of oil or other hazardous substances into or upon the navigable waters, adjacent shoreline, or contiguous zone, with the exception of those discharges permitted by the 1954 Convention, and in those cases where the discharge is judged to be not harmful. It requires that a person in charge of a facility notify the appropriate government agency of a prohibited oil discharge as soon as he is made aware of it, or face a criminal penalty of up to $10,000 fine and one year imprisonment, or both. The facility owners or operators are subject to a civil penalty of up to $5000. Factors in extenuation and mitigation as well as size of the business firm in question and the seriousness of the violation are considered during penalty assessment. The vessel owners or operators are liable to the United States for the actual clean-up costs up to an amount of $14,000,000 or $100 per gross ton of vessel, whichever is less. Willful negligence or misconduct has no maximum limit. Interestingly, willful negligence or misconduct in the case of an offshore facility is limited to a maximum of $8,000,000.

In addition to federal legislation, several states have wrestled with the liability issue. State-enacted legislation includes the Florida Oil Spill Prevention and Pollution Control Act, the Massachusetts Clean Waters Act, and the Delaware Coastal Zone Act.[36] State laws have had an extremely poor track record, however. Conflicts with federal law and jurisdiction, admiralty jurisdiction, and constitutionality have greatly hampered their effectiveness.

One last important consideration of the liability issue concerns the question of insurance.

There is a trade-off involved between the question of limited versus unlimited liability for injury and strict liability without fault. It directly involves the availability of insurance to shipowners. If insurance for an unlimited amount is available, there is likely to be less opposition to legislation imposing strict liability. At present the FWPCA calls for a limit of 14 million dollars or 100 dollars per gross ton on vessels and 8 million dollars on offshore and onshore facilities. A prime reason for these limits was the unavailability of insurance for more than these amounts.[37]

Current insurance requirements exist under the Federal Water Pollution Control Act but pertain only to vessels. They correspond to the limits of liability spelled out in the act. However, as demonstrated in the *Torrey Canyon* incident, the potential damage that can be incurred from an oil spill, or from the accidental release of hazardous materials carried in bulk, could easily exceed these upper limits. The question of insurance is highly relevant to any discussion of deepwater ports. Unfortunately the insurance question remains precisely that: a question with no obvious answers. Some workable alternatives to commercially available insurance have been fabricated, however:

But there are alternatives to commercial insurance as demonstrated by the Tanker Owner's Voluntary Agreement Concerning Liability for Oil Pollution (TOVALOP). A group of seven major oil companies formed a mutual insurance syndicate, the International Tankers Indemnity Association (ITIA). The terms of the agreement made the United States government a third party beneficiary with certain rights to recover for oil removal expenses, predicated on actual damage occurring or an immediate grave threat of damage. . . . If commercial insurance is limited, why not a supplementary self-insurance fund? The possibility of an oil spill is a risk of doing business and should not be arbitrarily limited.[38]

This concludes the description of liability, the second area of legal issues to be covered. The discussion turns now to the third and final area: ownership and control.

The discussion of TOVALOP as a feasible alternative by which oil companies can join together in an effort to meet unusually high insurance demands leads directly to the question of allowing oil companies to join together for the purpose of pooling resources to meet unusually high construction and operating costs. The cost of building and operating a deepwater terminal by any single firm may be prohibitive. For this reason and others, oil and chemical companies might decide to finance, build, and operate a deepwater terminal in one or more incorporated joint ventures. Joint ventures of this sort, however, raise the question of possible violations to the Sherman and Clayton acts.

Section 1 of the Sherman Act states:

Every contract, combination in the form of trust or otherwise, of conspiracy, in restraint of trade or commerce among the several States or with foreign nations is declared to be illegal.[39]

The issue raised here is whether or not a joint venture on the part of oil companies would constitute a "combination." Indeed, it is possible that they would be considered as such. In *Associated Press v. United States,*[40] in finding that a joint venture on the part of twelve hundred newspaper members comprised a combination, the court stated, "The Sherman Act was specifically intended to prohibit independent business from becoming 'associates' in a common plan which is bound to reduce their competitors' opportunity to buy or sell things (in) which the groups compete."[41]

A second issue is whether this combination constitutes an unreasonable "restraint of trade." Several cases have served to define the intent of the law with respect to the term "restraint of trade." *Standard Oil Company v. United States, United States v. American Tobacco Company,* and *United States v. Terminal Railroad Associates of St. Louis* are a few such cases.[42]

In *United States v. Terminal Railroad Associates of St. Louis,* for example, the court made it clear that combinations in and of themselves do not constitute restraints of trade. The court found it perfectly acceptable for companies to combine for the purpose of controlling or acquiring terminals for their common but exclusive use, *under ordinary circumstances.*[43] Deepwater terminals, how-ever, as in the St. Louis case, will probably not fall under the category of *"ordinary circumstances."* According to a study conducted by Nathan Asso-ciates, Inc., for the Institute of Water Resources:

However, actions taken by governments at the state and local levels to prevent the construction of new refineries and/or deepwater port facilities for crude petroleum have introduced a great element of uncertainty regarding the location of additional refinery required to meet East Coast needs. These actions raise questions as to whether such capacity will be located in this region at all.[44]

In other words, deepwater terminal financing, construction, and operation are constrained not only by economic cost considerations, but by political factors as well. Even if a terminal were a relatively inexpensive item to build and operate, the fact that political considerations limit the actual number that could ever be built gives those companies sharing ownership and control a sharp competitive edge over those that do not. There can be no question that based upon economies of scale alone, terminal owners and operators would have a great economic advantage over nonterminal owners and operators.

Section 7 of the Clayton Act as amended in 1950 states:

No corporation engaged in commerce shall acquire, directly or indirectly, the whole or any part of the stock or other share capital and no corporation subject to the jurisdiction of the Federal Trade Commission shall acquire the whole or any part of the assets of another corporation engaged also in commerce, where in any line of commerce in any section of the country, the effect of such acquisition may be substantially to lessen competition or to tend to create a monopoly.[45]

Subsequent adjudication of alleged violations of Section 7 of the Clayton Act have greatly shaped the present interpretation of this law. While the Sherman Act is primarily concerned with accomplished facts in restraint of trade such as monopolies and attempts to monopolize, the Clayton Act is concerned with the future impacts of current mergers, joint ventures, and other similar activities as they contribute to the lessening of competition or tendency to monopolize.[46] However, it is possible that the Clayton Act may not become an issue in a joint venture to finance, build, and operate an offshore deepwater terminal.[47]

The joint venture to build and operate a terminal would not result in the creation of a competitive company in competition with the constituent companies or with other deepwater terminals. For this reason it could be interpreted that there is no substantial lessening of competition nor a tendency to monopolize as defined by Section 7 of the Clayton Act.

It must be concluded that the Sherman and Clayton acts will have an impact on the final ownership and control structure of any offshore deepwater port in the United States. Sections 1 and 2 of the Sherman Act will probably have the greatest net effect on the final contractual terms adopted by the controlling companies and the policy formulated to accommodate those companies that are not contributing directly to the terminal's ownership or control.

Political Issues

It is extremely difficult to give short, concise explanations and descriptions of the political tug-of-wars that have transpired over the offshore deepwater terminal question. This difficulty is incurred for a variety of reasons. For example, many critical conversations and other important contributing factors have not been recorded or documented. Those that have been often contain information that is more emotionally charged than factual. Further, the identification of the real motivational factors underlying certain key rationales, decisions, etc., are often obscured, and assessments as to why certain things were said or done is left to speculative conjecture.

Yet it is extremely important that the political issues be addressed. Most of the confusion and uncertainty surrounding the planning for, and development of, deepwater terminals in this country originates in the political arena. Therefore, an understanding of the political activities that have occurred is an essential prerequisite for any understanding of deepwater terminal development or nondevelopment issues.

The question raised at this point is how to go about the description of political activity in a short, concise manner. One practical method is to analyze the laws that have been enacted in response to proposed deepwater terminal development. This approach is valid for two reasons. First, because there is no administrative process available at the federal, state, or local level for handling deepwater terminal siting, plan approval, standard development, regulation, etc., the only recourse for interested and/or involved parties has been through legislative chambers and/or the courts. Second, laws can be viewed as an output of the political process. They represent a recorded, extremely well-documented source of information. An analysis of these laws and the supportive background material for them, therefore, will provide insight into the real substance of the issues, political maneuverings, decisions, etc., that led to their enactment.

A second, less practical method is to conduct an exhaustive literature search and select a few representative passages, articles, etc., that best portray the political issues articulated in the bulk of the researched material. This method is somewhat less than practical because it falls prey too easily to the personal biases of reader and writer alike. The intent of the original works, the selection of "representative" passages presented, and the meaning to be conveyed are all exposed to subjective interpretation. Nevertheless, having expressed this cautionary note, both methods are used in the following discussion.

The Rivers and Harbors Act of 1899, the National Environmental Policy Act of 1969, the Federal Water Pollution Control Act (with 1970 and 1972 amendments), and the Coastal Zone Management Act of 1972 have all had an impact on deepwater terminal development in the United States. Fortunately for this discussion, one piece of legislation captures the spirit of the environmental movement that swept the United States in the late 1960s. An examination of this legislation—the National Environmental Policy Act of 1969 (NEPA)—provides a representative overview of all the political issues prevalent in laws subsequently passed.[48]

Passage of the National Environmental Policy Act was the most significant advance of the environmental movement. Several members of the United States Senate, notably senators Henry Jackson and Edmund Muskie, had proposed legislation that would establish U.S. policy on environmental issues. Jackson's proposal was supplemented with Muskie's proposal to form the basis of NEPA.[49]

The original intentions of NEPA's founders were to: (1) provide a statutory foundation for environmental litigation; (2) provide a reference basis for environmental concerns; and (3) provide procedures for handling present and future environmental issues.[50] The act provides for environmental communications links between government agencies; it provides a new basis for citizen litigation; and it directs its efforts toward incorporating environmental interests into the decision-making processes of governmental agencies.[51] The act further established the Council on Environmental Quality, which serves in an advisory capacity to the president.[52]

At the time of NEPA's passing, it was viewed as a relatively unimportant piece of legislation. Most felt that the act would not be passed, because it had no economic, social, or political constituencies. Yet after a single day of hearings the act was passed by an overwhelming voice vote of 372 to 15.[53] NEPA has had a significant impact on all major projects, like deepwater terminals, that have been planned or proposed since its enactment. NEPA's inherent strength lies with Section 102 of the act, which "provides a mechanism for significant reform in Government decision making."[54] While NEPA may or may not have reformed government decision-making processes, it certainly has made environmental considerations an important decision factor. The mechanism referred to makes its presence felt through: (1) required agency submittals of detailed environmental impact statements; and (2) making agency decision procedures available for public evaluation.[55]

It should be obvious that NEPA was not passed specifically in response to the deepwater terminal issue. But it was passed to handle *environmental issues* like those raised by deepwater terminal proposals. This realization provides a key to understanding the issues that make the deepwater terminal question so controversial. But before initiating an analysis of those issues and describing how they seem to fit together best conceptually, it is necessary to present the summary of researched material and the discussion of socioeconomic factors, in an effort to round out the picture of political activity that has affected the development of deepwater ports. This should shake loose the few remaining pieces in the deepwater terminal issue. Only after all the pieces are spread out on the table for analysis can the real challenge of deepwater ports be fully appreciated.

The discussion of political issues will turn now from legislation to focus on the voluminous literature. Article titles convey a great deal of information as to the methods used by "environmentalists" in asserting their position politically.[56] The titles alone are almost as enlightening as the articles themselves. The common thread carried throughout these and other articles appears to be, "How to get visibility for environmental issues." Sax makes the point quite emphatically that individuals should push to get precedent-setting environmental cases into court, regardless of their chances of winning.[57] Other articles, describing how to be an effective lobbyist or describing the use of multimedia confrontation for bringing about no-win agency proceedings, portray a different method of pursuing the same end. The realization as to what that end is provides a second key to understanding the issues that have created the controversy over the deepwater terminal.

Further clues are obtained through literature searches in the area of coastal-zone management. As stated previously, a deepwater terminal constitutes but one use among many contenders for the scarce resources of the coastal zone. Like many other contenders, i.e. beach and resort areas, deepwater terminals are unique in that they have nowhere else to go. They have no other alternatives except for those resources that are present in the coastal zone. This site-potential limitation is critical to deepwater terminal issues.

Over 50 percent of the population of the United States lives in a coastal zone, and the percentage is increasing.[58] The "northeast megalopolis" extends from Virginia to Maine and includes the cities of Portland, Boston, Providence, New York, Philadelphia, Baltimore, Washington, and Norfolk.[59] Over 45 million people, which is over 20 percent of the nation's population, live in this megalopolis comprising under 2 percent of the nation's land.[60] The population in the megalopolis is growing at a rate of 2 percent per year.[61]

As a resource, fishermen harvest nearly 5 billion pounds of fish in the coastal zone in a year. Estuaries alone are twice as productive, or more, than agriculture, per unit area.[62] As mentioned in the previous chapter, recreation in the coastal state of New Jersey, which has the largest refining capacity of any state in the nation, more than doubled the contribution to state revenue brought in by oil refining.[63]

The coastal zone is a receptacle for wastes, a source of raw materials like oil, sand, and gravel, and other valuable minerals, and a most desirable location for residential and industrial development. The Commission on Marine Science, Engineering and Resources notes that 7 percent of the nation's important estuarine areas were lost as a result of housing development during the past twenty years.[64]

As the migration into the coastal zone continues, more land will be required for housing at the sea's edge and in from the immediate shoreline. The piecemeal, unplanned, uncontrolled, unregulated system that has been used in many areas cannot continue. Growth must be planned for where it is likely to occur, adequate services must be planned in advance, and population densities must be regulated so as not to exceed the natural and service capacity of the area.... Effective use of planning and management techniques can reduce or eliminate uncontrolled urban sprawl, achieve higher aesthetic standards, and maintain economic balance. In addition, regionalization should be given serious consideration in planning the future of an urbanized coastal zone.[65]

Summarizing the issues touched upon in the last four paragraphs, deepwater terminals must be viewed as a proposed competing user of the scarce resources of the coastal zone. Due to population densities and economic composition, the coastal-zone areas are politically and economically sensitive. The coastal-zone areas are already the most competitive in the nation, without having a deepwater terminal contender thrust between existing competitors. Proposed deepwater terminals will be subjected to closer and closer scrutiny in terms of their ability to conform to emerging coastal-zone plans to satisfy developing control agencies, and to meet standards being set in proposed regulations.

Socioeconomic Issues

Chapter 4 presented the environmental considerations that have been addressed as they related to offshore-terminal structures, choice of transfer methods, vessel and terminal operations, and oil pollution. This presentation was conducted in this manner because it was representative of the issues discussed in reports, studies, analyses, etc., on deepwater ports from 1968 to 1973. Late in 1971 the emphasis on the ecological impacts of deepwater ports began to shift. An Arthur D. Little report on foreign deepwater port development found that it was the secondary industrial build-up that resulted in the greatest environmental and ecological change.[66] In fact, the correlation between secondary build-up and port location is so direct that France has utilized the pattern to counterbalance the preponderant economic weight of Paris as well as to spread out populace from congested to uncongested areas.[67] By selectively siting its deepwater ports, France has been able to selectively plan and expand urban and industrial development in formerly underdeveloped areas.

This report and others that followed addressed the issue of "secondary build-up" and its impact upon the quality of life. From 1971 on, the terms "secondary build-up" and "quality of life" occur more and more frequently. Deepwater terminal advocates were beginning to focus their attention on problems at the heart of the controversy.

Earlier in this chapter the statement was made that although "fear" might be a motivating factor in rallying opposition to deepwater port development, the role played by "fear of oil pollution" could conceivably be a small one. Several reliable sources were quoted to substantiate that claim. Continuing with one of those sources:

We have watched the functions of the oil companies, not only in New Jersey, and New Jersey probably is the best place to look at what an oil company can do. We are the largest refining state in the whole union of petroleum products, and we have watched the refining of petroleum products just about destroy all of the streams in the northern part of our state. . . .

Although New Jersey is the largest refiner of petroleum products in the whole of the United States, refining petroleum is not the largest industry in New Jersey, nor is it close to being the largest industry in New Jersey. New Jersey's largest industry, which is predominantly the area that I represent, the southern part of the state, happens to be the resort business, and the resort business by a ratio of almost two to one outranks oil refining in the state of New Jersey.

The county I live in is typical of what happens in New Jersey in the summertime. The population of Cape May County increases ten times every summer—ten times as many people in the summer as we have in the winter. They come for one big reason: because of the beaches that we have and because of the resort climate. And we want to keep it that way. This is our biggest industry, not oil refining. This is one of the things we would like to make very clear.[68]

Thus, socioeconomic disruptions were the motivating factors behind New Jersey's objections to the siting of a deepwater terminal in their coastal waters. The same conclusion can be drawn for other states. A quick rereading of most of the statements made in opposition to deepwater terminals as presented in this and the previous chapter reveals the presence of this underlying motivation.

As previously stated, an attempt to build an oil terminal and refinery at Machiasport, Maine, was blocked by generated public concern over the accuracy of promised future *economic* benefits and over the possible effects that oil pollution could have on existing *fisheries* and *tourism*. Governor Peterson of Delaware stated (p. 74) that he would oppose a plant that was 100 percent pollution free, and that Delaware's shoreline would be reserved for tourism, recreation and other *compatible* industrial uses. Such sentiments are expressed again and again in reports, research papers, public-hearings proceedings, etc. (It is interesting to note that these socioeconomic concerns are invariably categorized as environmental concerns.)

It is even more interesting that the terms "environment" or "environmental" are never defined. In view of the ambiguity of the word "environment" this realization is significant. There are home environments, work environments, corporate environments, social environments, natural environments, etc. It would be very simple for a socioeconomic environment to be described as "quality of life." The quality of life is also a logical extension of environmental concern. The loose usage of the word "environment" would be enough to allow socioeconomic issues to function effectively under the guise of environmental issues. In this case, early environmental impact statements that addressed the impacts that a deepwater terminal would have on the natural or ecological environment would be missing the major issues.

The major issues are those which relate to socioeconomic disruptions that might occur in the vicinity of a deepwater terminal. These are the issues that stir the opposition. The presence of a deepwater terminal could, and most probably would, shift the economic basis of the selected community. Changes in property values, tax rates, zoning, labor markets, residential and commercial properties, etc., would constitute the on-shore impacts that would have to be addressed in the environmental impact statements. Such a task is beyond current institutional mechanisms.

Conceptualizing the Issues

If the "environmentalists v. economists" model is so misleading in any attempt to conceptualize the major issues, what paradigm is available to replace it? If a dichotomy is desired, then a better replacement would be "national v. local." This approach produces a more functional, more representative model of the issues. It would be far more useful to assess national economic gains or losses against local economic gains or losses and national environmental gains and losses against local environmental gains and losses. This is far more logical than pursuing the present system, which attempts to weigh economic gains against environmental impacts. The former are comparable, the latter are not.

If "national v. local" is a superior means of viewing the controversy, why hasn't it been used? The answer to this question is not readily apparent. Most probably, however, it is because of the fact that the lack of comparability has served those who oppose deepwater ports as well as environmentalists. Regardless of their motivation, the ends pursued by environmentalists and deepwater port opponents have been similar. For the former, it has been, "Stop bulk oil carriers (i.e., the deepwater terminal pipe) from further degrading the quality of our oceans!" For the latter, it has been, "Stop the deepwater terminal pipe from landing in my back yard and degrading my quality of life!" Quite obviously, the common thread is citizen concern, and that concern has blocked deepwater terminal development. Several mechanisms have been available to environmentalists to express this concern, and all have been used. Summarizing these mechanisms in terms of a national v. local construct is somewhat enlightening.

First, since the complexity of issues which revolve around financing, building, and controlling deepwater terminals requires laborious, time-consuming efforts in and of themselves, questions requiring extensive research and several years of effort raised at strategic times within the executive branch of government have served to slow and postpone the development of deepwater ports. This tactical approach has also served to raise uncertainty, increasing the risks associated with deepwater terminal planning proposals, making deepwater terminals a decreasingly desirable venture for which to solicit investment capital.

A second approach has been to take positive action within the legislative branch of government. As the reader may have noticed, environmental issues and legal issues are inextricably intertwined. Obviously, this is not coincidental. Since the mechanisms for seeking administrative relief were nonexistent within the executive branch of government, environmentalists were able to make inroads directly within the legislative and judicial branches (i.e., the traditional process of exhausting administrative remedies first was totally circumvented). Deepwater terminal opponents have therefore benefited by joining environmentalists in supporting strategic pieces of legislation and specific, precedent-setting court cases. The resultant laws and precedents, in turn, have served to slow deepwater terminal development, by requiring advocates to substantiate and justify their proposed plans at every step of the decision-making process.

A third device has been to increase the public's awareness of environmental issues. Advocates have been able to respond quickly in this area also, producing what has been termed the "mass-media confrontation." While deepwater terminal advocates presently dominate the mass media, the earlier efforts of environmentalists in this area no doubt paved the way for the key pieces of legislation passed in the late sixties and early seventies in support of their cause. This conclusion is further substantiated by the realization that some major environmentalist groups now endorse deepwater terminals as being environmentally desirable in comparison to present oil transfer methods. This stand has rendered terminal opponents incapable of continuing a mass-media confrontation over the environmental aspects of deepwater terminals.

A fourth mechanism has been through application of political pressure on local and regional elected officials. This method is considerably different from the legal and judicial strategies mentioned earlier. Legal strategies attempt to institutionalize environmentalism; political strategies attempt to pressure the status quo. Thus, deepwater terminal opponents have attempted to make their respective geographic locales exceedingly undesirable places to site a terminal or land a pipeline. This pressure becomes apparent in terms of local economic policy, latitude offered in zoning variances, state and municipal agency cooperation, etc.

Looking at the total picture, environmental and economic issues and concerns are critical factors at both the local and national level. Figure 5-2 diagrammatically depicts the relationship between economic and environmental issues in terms of a national versus local construct. Figure 5-1 is shown for purposes of comparison, to depict the "economists vs. environmentalists" dichotomy which most of the literature addressed from 1968 to 1973.

The conceptual model in Figure 5-2 is clearly superior in that all other issues impacting upon deepwater terminal development in this country can be

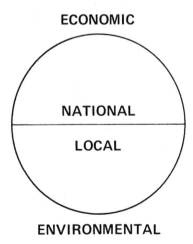

ECONOMIC

NATIONAL

LOCAL

ENVIRONMENTAL

Figure 5-1. Common Picture of "Economists vs. Environmentalists" Dichotomy

NATIONAL

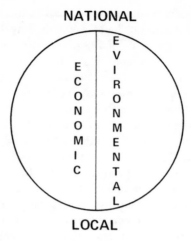

LOCAL

Figure 5-2. Revised View of "Economists vs. Environmentalists" Dichotomy

graphically displayed upon it. For example, social factors prevalent at the local levels have been present in both the economic and environmental sectors. Some were socioeconomic and some were what could be termed socioenvironmental. Taken in the aggregate, they were labeled as quality-of-life issues.

While some legal actions were taken at the state level, most of the successful legal sorties and legal questions occurred at the national level. Both impacted upon the economic and environmental sectors. Legal issues in the economic area concentrate on questions of antitrust, labor union factions, liability and insurance, etc. In the environmental area, they are manifest in specific legislation like the Coastal Zone Management Act, the Federal Water Pollution Control Act, the National Environmental Policy Act, and other key precedent-setting adjudication.

Political factors are present at both the national and local levels, and act as a buffer between legal, social, technological, economic and environmental issues and national and local priorities and concerns. Figure 5-3 depicts this final conceptual tool. It is offered as a method for viewing the vast diversity of information contained in this book.

In summary, this chapter has addressed most of the major issues, other than strictly economic or environmental, that have influenced the development of deepwater terminals or the landing of a deepwater terminal pipe in this country. The issues addressed were then used to develop the model shown in Figure 5-3. The information presented in Chapters 1, 2, and 3 are represented in the economic sector of the economic hemisphere. The information presented in Chapter 4 is represented in the environmental sector of the environmental hemisphere. Chapter 6 discusses the aspects of federal government regulation

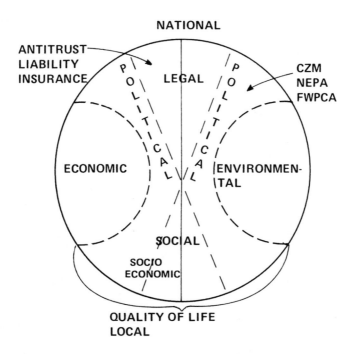

Figure 5-3. Conceptual Model of Multiple Issues Involved in Economic/Environmental Concerns

present in the political sector of the model, and Chapter 7 adds the labor dimensions to the socioeconomic sector. This model, in essence, is the conclusion of this chapter.

6

The Role of the Federal Government[a]

The United States is one of the few major industrial nations that does not have a deepwater terminal. While Mother Nature has played a role in this outcome by giving the U.S. a general lack of natural deepwater harbors, other factors have been more important in bringing about this result. A key factor is the traditionally weak role played by the U.S. federal government in port planning and development. Foreign governments typically play a more active role.

Foreign Experience

A study performed by Arthur D. Little, Inc., for the Institute for Water Resources of the Army Corps of Engineers surveyed the role of several foreign governments in port planning and development.[b] In each country, some kind of organized coordination among local, regional, and national governments exists, and port planning is performed on a long-term basis covering a ten-to-thirty-year range. The national government typically plays the leading role, and the planning cuts across municipal and subregional boundaries.

In the Netherlands the national government, the provinces, the regional government, the community, and the port authorities are all brought into the initial planning. Port planning in France, referred to in Chapter 5, makes use of regional planning teams reflecting not only diverse interest groups but also widely varied disciplines such as those of architects, urban specialists, sociologists, economists, and industry and transport specialists.[1]

The concern of foreign governments for port development is reflected in part by Figure 6-1, which shows merchandise exports for several nations as a percent of the gross domestic product. In the U.S., merchandise exports contribute only 4.5 percent of the gross domestic product, considerably less than several other nations whose merchandise exports contribute between 10.0 percent and 46.5 percent of their gross domestic products.

[a]Material in this chapter relies on material contained in Henry S. Marcus, James E. Short, John C. Kuypers, and Paul O. Roberts, *Federal Port Policy in the United States* (MIT, 1975).

[b]See Arthur D. Little, Inc., *Foreign Deep Water Port Developments*, sponsored by the Institute for Water Resources, Department of the Army, Corps of Engineers, IWR Report 71-11, December 1971, for descriptions of examples of governmental port planning in foreign countries.

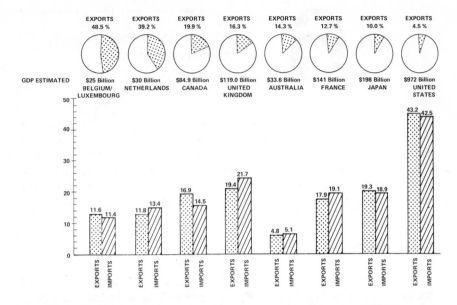

Figure 6-1. Merchandise Exports as a Percent of Gross Domestic Product, 1970. Source: Arthur D. Little, Inc. *Foreign Deep Water Port Developments*, sponsored by The Institute for Water Resources, Department of the Army, Corps of Engineers, IWR Report 7-11, December 1971.

Fragmentation of United States Power

In order to understand federal port policy in the United States, one must be aware of the federal power structure as it relates to ports. A myriad of federal government organizations are involved in port development and operations in the United States, as shown in Table 6-1. Power to influence port actions is remarkably fragmented among these federal bodies. In fact, certain regulatory powers may come as a surprise to the general layman. For example, the Secretary of Defense (through the Army Corps of Engineers) is responsible for determining the commercial adequacy of ports, while the Secretary of Commerce (through the Maritime Administration) is responsible for mobilization of ports in time of war. One might have easily expected these roles to be reversed.

If U.S. deepwater ports consist of offshore terminals which handle only bulk commodities, some of the federal organizations dealing with other types of cargo and passengers will not be applicable (i.e., Immigration and Naturalization Services, Bureau of Animal Husbandry, etc.). Table 6-2 identifies thirty-three topics germane to the design and operation of deepwater ports, and shows the federal agencies concerned with these areas.

Table 6-1
Federal Organizations Involved in Port and Harbor Development

Atomic Energy Commission (AEC)

Council on Environmental Quality (CEQ)

Department of Agriculture
    Bureau of Animal Husbandry
    Bureau of Entomology and Plant Quarantine

Department of the Army
    Army Corps of Engineers
        Board of Engineers for Rivers and Harbors

Department of Commerce
    Economic Development Administration
    Maritime Administration
    National Oceanographic and Atmospheric Administration
        Office of Coastal Zone Management
        National Ocean Survey
        National Marine Fisheries Service
        National Weather Service
        Sea Grant Program

Department of Defense (Departments of the Army and the Navy)

Department of Health, Education and Welfare (HEW)
    Public Health Service

Department of Housing and Urban Development (HUD)
    Housing and Home Finance Agency
        Community Facilities Administration
        Urban Renewal Administration

Department of the Interior
    Geological Survey
    Bureau of Land Management
    Bureau of Sport Fisheries and Wildlife
    Office of Land Use and Water Planning
    Office of Water Resources Research

Department of Justice
    Immigration and Naturalization Service

Department of Labor
    Occupational Safety and Health Administration

Department of the Navy
    Oceanographic Office

Department of State

Department of Transportation (DOT)
    The United States Coast Guard

Department of the Treasury
    Bureau of Customs
    Internal Revenue Service

The Executive Offices of the President
    Council of Economic Advisers
    Domestic Council
    Office of Management and Budget (OMB)

Table 6-1 (cont.)

Environmental Protection Agency (EPA)

Federal Communications Commission (FCC)

Federal Energy Administration (FEA)

Federal Maritime Commission (FMC)

Federal Power Commission (FPC)

Federal Trade Commission (FTC)

General Services Administration (GSA)

Interstate Commerce Commission (ICC)

National Aeronautics and Space Administration (NASA)

Office of Economic Opportunity (OEO)

Smithsonian Institution

United States Congress

United States Postal Service

United States Supreme Court

Water Resources Council

Source: Henry S. Marcus, James E. Short, John C. Kuypers, and Paul O. Roberts, *Federal Port Policy in the United States,* (MIT, 1975).

Territorial Jurisdiction[c]

For deepwater terminals within the territorial seas of the United States, existing regulations cover most, if not all, of the topics covered in Table 6-2. However, most proposed superports are to be located beyond U.S. territorial seas, and problems arise in these instances.

The Territorial Sea Convention establishes the breadth of the territorial sea as extending from the low-water line along the coast to a distance of three nautical miles into the sea. Consequently, the United States has full sovereignty over water along its coast within these limits. In the 1953 Submerged Lands Act, the federal government granted to the states its full legal jurisdiction over the seabed up to the three-mile limit, reserving only its traditional functional responsibilities for commerce and navigation. In most instances federally delegated jurisdiction remains at the state level.[2]

Until recently the only relevant U.S. legal jurisdiction beyond its present territorial sea limit of three miles concerned the exploration and exploitation of natural resources in the seabed and subsoil to a depth of 200 meters (well beyond any potential deepwater port terminal), which it obtained under the Continental Shelf Convention and its national equivalent (the Outer Continental Shelf Act of 1953, as amended). All existing deepwater ports in the world

[c]For further information on this topic the reader is referred to the following publications: *Handbook For Offshore Port Planning,* edited by Patrick J. McWethy and Stewart B. Nelson, Marine Technology Society, Washington, D.C., 1974; *Perspectives on Oil Refineries and Offshore Unloading Facilities, Proceedings, The Fourth New England Coastal Zone Management Conference,* edited by Mary Louise Hunter, Durham, N.H., May 13-14, 1974.

Table 6-2
Deepwater Port Topics

Topic	Department	Agencies Involved
Geology	DOI	Geological Survey
Ocean currents	DOC	National Oceanic and Atmospheric Administration
	DOD	Navy
		Corps of Engineers
Salinity/temperature	DOC	National Oceanic and Atmospheric Administration
	DOD	Navy
		Corps of Engineers
Climate	DOC	National Oceanic and Atmospheric Administration
Biologic life	DOI	Bureau of Sport Fisheries and Wildlife
	DOC	National Oceanic and Atmospheric Administration
Navigation	DOT	U.S. Coast Guard
	DOD	Corps of Engineers
Commercial fishing	DOC	National Oceanic and Atmospheric Administration
Outdoor recreation	DOI	Bureau of Outdoor Recreation National Park Service
Effects on shoreline	DOD	Corps of Engineers
	DOI	Land Use Planning
	DOC	National Oceanic Atmospheric Administration
Wave size	DOC	National Oceanic and Atmospheric Administration
	DOD	Navy
		Corps of Engineers
Pipeline construction	DOT	Office of Pipeline Safety
	DOI	Geological Survey
		Bureau of Land Management
	DOD	Corps of Engineers
Dredging/filling	DOD	Corps of Engineers
	DOI	Bureau of Sport Fisheries and Wildlife
	DOC	National Oceanic and Atmospheric Administration
Pipeline safety	DOT	Office of Pipeline Safety
	DOI	Geological Survey
Platform safety	DOI	Geological Survey
	DOT	U.S. Coast Guard
Vessel operations safety	DOT	U.S. Coast Guard
Platform design and construction	DOI	Geological Survey
	DOD	Corps of Engineers
	DOC	Marad

Table 6-2 (cont.)

Topic	Department	Agencies Involved
Deepwater port operations	DOT	U.S. Coast Guard
	DOI	Geological Survey
	DOD	Corps of Engineers
Navigation aids	DOT	U.S. Coast Guard
Navigation operations	DOT	U.S. Coast Guard
	DOD	Corps of Engineers
Pipeline construction (land)	DOD	Corps of Engineers
	DOI	Geological Survey
		Land Use and Water Planning
Siting	DOI	Bureau of Land Management
		Geological Survey
	DOD	Corps of Engineers
Zoning for land installations	DOI	Office of Land Use and Water Planning
	DOC	National Oceanic and Atmospheric Administration
Pollution (air and water)		Environmental Protection Agency
	DOT	U.S. Coast Guard
	DOI	Geological Survey
Ocean dumping	DOD	Corps of Engineers
	DOI	Bureau of Sport Fisheries and Wildlife
	DOC	National Oceanic and Atmospheric Administration
Law enforcement (civil/criminal)	DOI	Geological Survey
		Bureau of Land Management
	DOD	Corps of Engineers
	DOT	U.S. Coast Guard
		Environmental Protection Agency
Tariff-user rates	Inter State Commerce Commission	
Fee schedule	DOI	Bureau of Land Management
	DOD	Corps of Engineers
	DOT	U.S. Coast Guard
Maritime technology	DOC	Marad
	DOI	Geological Survey
	DOT	U.S. Coast Guard
Bonding	DOI	Geological Survey
	DOD	Corps of Engineers
International careers	Department of State	
	DOD	
National security	DOD	
Economic analysis	DOI	Bureau of Land Management
	DOC	Marad
	DOD	Corps of Engineers

Table 6-2 (cont.)

Topic	Department	Agencies Involved
Environmental concerns		Council of Environmental Quality
		Environmental Protection Agency
	DOC	National Oceanic and Atmospheric Administration
	DOI	Geological Survey
		Bureau of Land Management
		Bureau of Sport Fisheries and Wildlife

Source: Background material for Senate Bill S. 1751.

located beyond a nation's territorial sea are directly related to exploitation of petroleum or natural gas from the Continental Shelf, and clearly fall within the permissible activities of the convention. However, United States offshore port development is presently contemplated entirely for petroleum imports, unrelated to resource extraction from its Continental Shelf. Since no other international conventions or laws treat deepwater ports, a study sponsored by the Corps of Engineers concluded in June 1973 that "apparently no legal regime at any level of government now applies to facilities beyond the United States territorial sea limit of three nautical miles."[3] The report concluded that new legislation was necessary to establish a basis for, and to assert, United States legal jurisdiction over facilities in international water beyond the territorial sea and to authorize the licensing of the construction and operation of such deepwater ports.[4] However, it took until December 1974 for Congress to pass such legislation, the Deepwater Port Act of 1974.

It is expected that the Law of the Sea conferences will eventually settle this issue on an international scale. However, progress in this area has been extremely slow to date.

Administrative Procedures

Traditionally, port development generally involves dredging operations by the United States Army Corps of Engineers. Such dredging of harbors and channels is done out of congressionally budgeted funds, with no direct cost to the affected port. However, the administrative procedures involved with such dredging projects (new construction projects rather than maintenance of earlier projects) are complex and extremely time consuming.

Testifying in June 1971 before the Subcommittee on Investigations and Oversight of the House Public Works Committee, Major General Frank Hoisch of the Army Corps of Engineers estimated that it would take seventeen years and eleven months from the initial authorization of a study to completion of construction in a civil works project.[5] Figure 6-2 shows the breakdown of time required for each phase of the overall approval process.

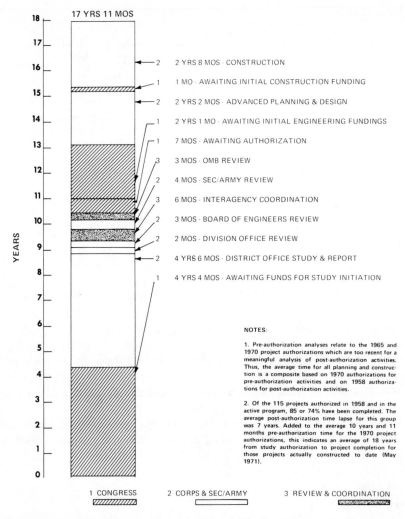

Figure 6-2. Corps of Engineers Analysis of Average Time for Planning and Construction of Civil Works Projects (May 1971 Status). Source: Authorization and Appropriation Processes for Water Resource Development, Cornell University, 1972.

The types of deepwater ports most likely to be built in the near future would consist of single-point moorings that require no dredging and no federal funds for construction. Consequently, the administrative procedures studies by the Corps of Engineers to justify federal funding and the congressional authorization of funds for studying, planning, and construction can be eliminated. However, even the process of obtaining a work permit from the Corps of Engineers for privately funded construction in United States navigable waters can be time consuming.

During congressional hearings on deepwater ports in 1973, General James V. Cross, Executive Director of the State of Texas Offshore Terminal Commission, presented the chart shown in Figure 6-3, which describes the procedure for obtaining a work permit from the Corps of Engineers.[6] Twenty-one agencies and six other groups are presented on the chart, and many of the steps indicated may have to be done several times before obtaining final approval. In the opinion of General Cross:

Such delaying procedures could hold up development of a deepwater port facility indefinitely. Accordingly, to avoid such delays, legislation authorizing permits for such facilities must contain specific provisions to require precise and expeditious handling of applications by a single Federal agency, and that agency should be given the necessary muscle to insist that expeditious handling be accorded each application by all agencies involved in the permitting/licensing process.[7]

The Deepwater Port Act

While it became apparent to the Congress that new legislation was necessary both to handle licensing and to settle jurisdictional problems, discussion of such legislation dragged on for two years. Unfortunately, the fragmentation of power between congressional committees and subcommittees as it relates to port planning and development is similar to that of the federal agencies. Consequently, intercommittee rivalry played a key role in this delay.

On January 4, 1975, President Ford signed the Deepwater Port Act of 1974,[d] creating a licensing mechanism for the construction and operation of port facilities handling oil and natural gas imports and exports located beyond the territorial waters of the United States. The act does not apply to offshore terminals proposed to be built within territorial waters. Port facilities within three miles will be licensed in the usual method—under the Rivers and Harbors Act by the U.S. Army Corps of Engineers. The salient features of the bill are the following:

1. *Licensing Agency.* The bill states that the Department of Transportation will be the single agency for licensing the construction and operation of deepwater ports. (The Department of the Interior, which includes the Bureau of Land Management, the Geological Survey, the Bureau of Sport Fisheries and Wildlife, the Office of Oil and Gas, and the Bureau of Outdoor Recreation, had been considered as the lead agency in a different version of the bill.) DOT will also provide for "full consultation and cooperation with all other interested federal agencies and departments, and with any potentially affected coastal state, and for consideration of the views of any interested members of the general public [Sec. 5a]."

This legislation complements earlier congressional mandates to DOT. By virtue of earlier legislation, the Secretary of Transportation is charged with the responsibility for development of transportation policies and programs, and

[d]See the Deepwater Port Act of 1974, Public Law 93-627, 88 Stat. 2126.

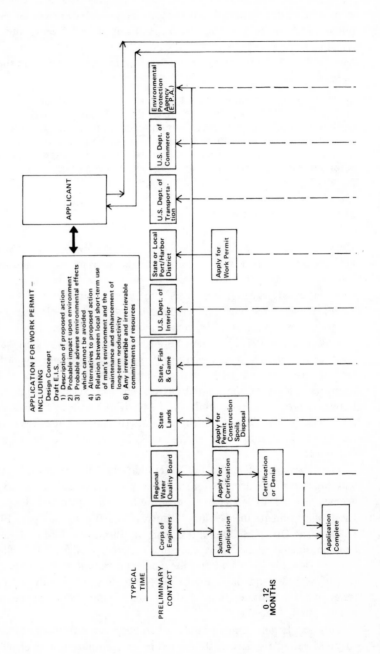

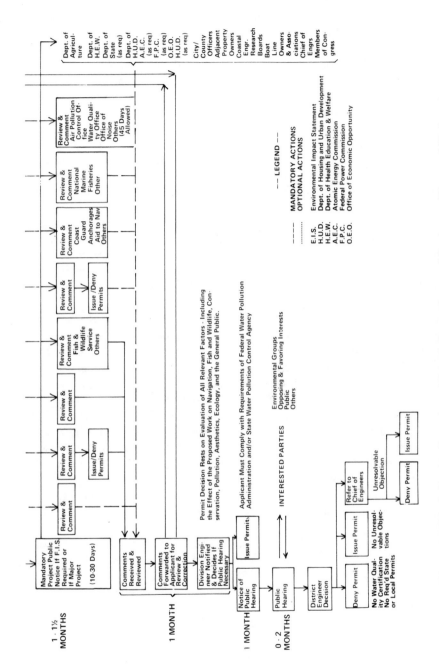

Figure 6-3. Work Permit Procedure: United States Corps of Engineers-River and Harbor Act.

therefore is concerned with the location of deepwater port facilities in a manner that is consistent with and supportive to the other elements of the transportation intrastructure. In addition, DOT contains the Office of Pipeline Safety as well as the Coast Guard, which is the primary maritime law-enforcement agency of the federal government. Coast Guard responsibilities within the ports of the United States also include merchant vessel safety, port safety, aids to navigation, and marine environmental protection, as well as search and rescue.

2. *Adjacent State Veto.* The bill provides adjacent coastal states with the right to veto any deepwater port proposed to be licensed under the act. An adjacent coastal state is broadly defined, and includes: (1) a state that is directly connected to the port by pipelines; (2) a state located within fifteen miles of the proposed port; and (3) a state threatened with a possible oil spill from the port.

3. *Procedure.* A timetable for action on a license of eleven months is established. The procedure involved includes application, environmental impact statement, hearings, and final action by all federal agencies. Applications to build ports in the same location are handled by a procedure designed to consider all applications for any one location. If all applicants qualify, the secretary is to issue a license according to a priority: (1) a state application; (2) an application by an independent terminal company; and (3) any other application.

4. *Environmental Review.* The Secretary of Transportation, together with the administrator of the EPA and the administrator of NOAA, is to establish environmental review criteria that shall be used to evaluate an application to build a deepwater port. The criteria are to include the full range of environmental concerns associated with deepwater ports.

5. *Antitrust Review.* Among the prerequisites to the issuance of a license is the requirement in section 7 for an antitrust review of the application by the attorney general and the Federal Trade Commission. Both agencies are to give the secretary an opinion as to whether issuance of the license would adversely affect competition, restrain trade, further monopolization, or otherwise create or maintain a situation in contravention of the antitrust laws.

6. *Common Carrier Status.* Existing statutes regulating the transportation of oil and natural gas in interstate commerce are made specifically applicable to a deepwater port.

7. *Navigational Safety.* The Department of Transportation (through the Coast Guard) is authorized to prescribe by regulation procedures to ensure navigational safety around and near the port. The DOT is further authorized to designate a safety zone around the port within which no uses incompatible with such port are to be permitted.

8. *Liability.* Strict liability for pollution damage caused by a discharge from the port itself or from a vessel within the safety zone is prescribed. The bill allocates liability among: (1) the licensee, up to fifty million dollars; (2) the owner and operator of a vessel, up to twenty million dollars; and (3) a deepwater port liability fund for all other proven damages (including clean-up

costs) not actually compensated by the licensee or the owner or operator. The fund, administered by the Secretary of Transportation, is created by a two-cents-a-barrel charge on oil until the fund reaches one hundred million dollars.

Economic Analysis

A further provision of the Deepwater Port Act provides that the Department of Transportation perform economic analysis to study the impact of offshore terminals on ports that have taken steps to plan for onshore deepwater terminals. In addition, the economic benefits of dredging an onshore deepwater port will be compared with building an offshore terminal. Such an analysis can be quite complex, since an onshore deepwater port may be handling break-bulk cargo, containers, and dry bulk, as well as the petroleum products which would be handled at the offshore terminal.

While the DOT has little experience in economic analysis of ports, other agencies are currently involved in such activities. The Corps of Engineers performs benefit/cost analyses on all proposed dredging projects using federal funds. The Maritime Administration has partially funded a regional port study of the United States Pacific Northwest, and tentatively plans such studies for other areas of the United States. Needless to say, there is overlap in the economic analyses done by these agencies. Unfortunately, however, there is no formal mechanism that coordinates the economic analyses made by these federal agencies. To complicate matters further, pending legislation calls for the expenditure of up to two million dollars of federal funds so that the Secretary of Commerce (apparently through the Maritime Administration) can "undertake a comprehensive study to determine the immediate and long-range requirements of public ports in the United States."

It would appear that the spending of taxpayers' funds for economic analyses of ports could be improved by stipulating in any new such legislation that the Corps, DOT, and the Maritime Administration should jointly perform any nationwide port study after determining a common data base, set of criteria, and methodology to be used in all their analyses. Such a legislative mandate would assure taxpayers that there would be little or no duplication of effort among these federal agencies. In addition, economic analyses at a local and regional level would be able to fit logically into the framework of a national port study.

Federal Role in Pollution Control

In addition to licensing deepwater ports, the federal government must also be concerned with controlling pollution of the seas. On the domestic level the federal government sets standards for vessels entering United States territorial waters. On the international level the government works through the Inter-Governmental Maritime Consultative Organization (IMCO), organized in 1948 as an advisory group to the United Nations, which had worked on a draft convention in 1962 to control pollution of the seas. In January 1967 IMCO's Maritime Safety Committee established a Subcommittee on Ship Design and Equipment, which soon after *Torrey Canyon* became one focal point for studies of antipollution measures. Some systems external to the oil tankship were considered, such as special training for tanker and port crews, port facilities for tank cleaning, and various navigational systems led by sophisticated vessel traffic control systems for harbors and areas of high potential hazard. Other studies concentrated on the tanker, i.e., the "integrity and reliability of the cargo containment system."[8]

Many of the antipollution measures external to the tanker—for example, a vessel traffic control system—represent benefits and costs to the sponsor nation, but many internal "containment systems" move across international lines with the tanker. If tankers of all registry were subject to the same additional antipollution measures, their costs in theory would rise a proportional increment. Yet if some tankers were subject to environment restraint, and others were not, the workings of a competitive market would tend to squeeze the operator, industry, and nation complying with restraint, while at the same time increasing the volume of the operator without the restraint. If countries have different standards and hence impose different costs on their tankers, the number of tankers registered in the country with effectively the lowest standards will rise, and the number of tankers registered in the country with the highest standards will decline, until the effective world standard would be the lowest standard set by an individual country of registry. Even before this theoretical limit would ever be reached the country of registry of tankers bearing costs not required of competitors would also suffer in trade balances, fleet availability in time of emergency, and a decline in employment in the domestic industry. At the same time trade would go to the country less concerned. It is apparent that international cooperation and agreement are necessary to overcome such inequities.[e]

The issue of providing antipollution measures in the international tanker market was one of the large issues dealt with at the Marine Pollution Conference, attended by representatives of seventy nations, concluded in London on November 2, 1973. Admiral C.R. Bender, Coast Guard Commandant and Vice Chairman of the U.S. delegation, stated that the "new convention can become a basis for Coast Guard enforcement and regulatory activity to control pollution from all vessels entering U.S. ports." The convention must now be ratified by governments before it becomes effective.[9]

---

[e]For a discussion of the dilemma of economic regulation in this international market see Louis K. Bragaw, "Environmental Policy Formulation in Competitive Tanker Markets," *Decision Sciences Northeast Proceedings*, vol. 3 (1974), pp. 31-34.

Russell Train stated that "we can be proud of the fact that the two years of international activity culminating in this convention followed an United States initiative, made in 1970, calling on the nations of the world to take action to end ship-generated marine pollution in this decade." Among the "major contributions" of the convention that will now go to the Senate are the following:

A requirement that newly constructed oil tankers over 70,000 tons deadweight have separate tanks for cargo and ballast . . .

Limitations on oil discharge from ships, including a complete prohibition on oil discharges within fifty miles of land.

Flexible amendment procedures to allow technical provisions to be kept up to date without traditional cumbersome treaty revision processes.[10]

Mr. Train went on to enumerate some "important new legal duties" that would be enacted to see that the "substantive regulations are respected," among them:

The Nation whose flag a ship flies must punish all violations by that ship.

In addition, nations must either punish violations by a foreign flag vessel which occur in their waters or refer them to the state of the flag for prosecution.

Nations must deny permission to leave their ports to ships which do not substantially comply with the convention's construction requirements, such as segregated ballast tanks, until these ships can sail without presenting an unreasonable threat to the marine environment.

Nations which ratify the convention must apply its requirements to the ships of nations which do not ratify, as necessary to ensure that those ships do not receive favorable treatment, thus preventing a nation from obtaining competitive advantages by not joining the convention.

A system for settling disputes which arise under the convention through compulsory arbitration.[11]

It is hoped that the ratification and enforcement of this convention will play a significant role in helping the United States to meet the challenge of deepwater terminals.

7

Labor Aspects of Deepwater Terminal Systems

Labor and Marine Technology

The marine petroleum transportation system can be designed to meet whatever energy transport needs are placed upon it. The economic, environmental, and onshore aspects of the system have been discussed. In an industry as intensely unionized as the marine industry, it is important to assess the possible impacts and reactions of labor to any changes made in the operating system.[1]

Organized labor has sometimes slowed the process of innovation in marine technology. Containerization provided the marine field with an efficient method of moving general cargo. The longshoring unions, realizing that containerization would mean a loss of jobs on the docks, held many strikes, trying to obtain what they felt was their fair share of the benefits of modern technology. The new task of driving tractor and trailer combinations within a marine terminal became the issue of a jurisdictional dispute between unions. This interunion rivalry caused temporary shutdowns of container terminal facilities. The introduction of the nuclear ship *Savannah* also was delayed by union problems.[a] Union issues arose concerning both the wage levels and jurisdiction of competing unions. This chapter will look at potential labor issues arising from the introduction of deepwater terminals.

The Role of Labor in Deepwater Terminal Systems

Adding deepwater terminals to receive supertankers represents a major change at the tanker-terminal interface. Existing loading docks will be replaced by a terminal system which will come into contact with the traditional shore-based oil unions, coastal platform unions, and the offshore marine unions. Figure 7-1 illustrates the relative roles of the refinery and shore-based unions, the marine unions, and the platform unions as they cover various segments of existing systems. Figure 7-2 illustrates possible pressures that could be exerted by various unions as alterations are made in the delivery system.

This chapter will consider the labor aspects of deepwater terminal systems from several vantages: (1) the terminal's operations and the unions involved;

[a]For further information see David Kuechle, *The Story of The Savannah* (Cambridge, Mass.: Harvard University Press, 1971).

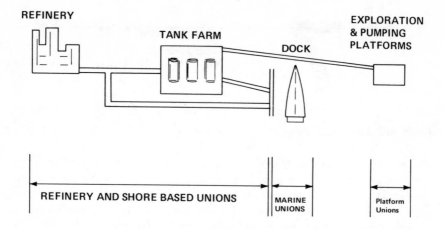

Figure 7-1. Relative Roles of Various Unions in Existing Terminal Systems

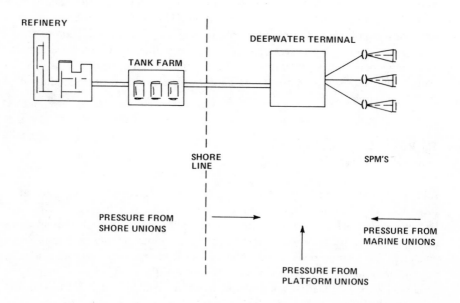

Figure 7-2. Relative Pressure of Various Unions in a Deepwater Terminal System

(2) the vessels; (3) the maritime unions; and (4) human factors. There are several manpower problems that should be examined. Problem areas include job-training requirements, job interfacing, and rivalries between existing unions. Any study of deepwater terminal systems without consideration of the labor aspects would be both limited and unrealistic.

Terminal Systems and Their Labor Unions

Currently most terminals consist of a loading dock, and are considered extensions of the refineries they serve. If a refinery is unionized, then the operations at the terminal or loading dock usually belong to the refinery union. In some instances the refineries are nonunion or organized by an independent union; however, the predominant union in this field is the Oil, Chemical and Atomic Workers (OCAW).

Oil dock unionization depends on the nature of the organization of the tank farm the dock serves. This is so because quite often oil dock personnel have primary job responsibilities elsewhere. A large percentage of oil docks serving tank farms, as opposed to refineries, are located in the remote areas of a port, and may or may not be organized at all. Oil docks are far from being the source of strength for the ILA, and most oil docks are not under ILA control. For these reasons terminal employees may be nonunion, or belong to the OCAW, an independent (a former company union), or some other union that is not normally associated with ships and ports.

As previously noted, current planning indicates that deepwater terminals along the Gulf Coast, and probably the East Coast as well, will consist of single-point moorings (SPMs) rather than artificial islands. The SPMs have the advantage of reduced manpower, and have streamlined vessel handling equipment. Vessels moor to the buoy under their own power with the assistance of a line-handling boat; they then use their own pumps and crew to discharge their cargo. Thus, the SPM system creates four separate and distinct labor groups: shoreside tank-farm employees, platform employees, SPM-related employees, and vessel employees. The first three will be considered here; the fourth will be discussed later.

Tank-farm employees will experience little difference in their duties and working environment as the system changes. This implies that recruiting and training will come from existing sectors of the industry. In all likelihood employees would organize under whatever union dominates the trade in the area of the farm.

It is expected that the SPM platform workers will consist of approximately forty to fifty maintenance men, cooks, messboys, dispatchers, and divers. The duties of the first three are essentially the same as exist on oil exploration and production platforms. For this reason wages and benefits would probably be tied to that industry. The possibility of union organizing among this group depends upon the level of activity on the exploration platforms. Many of the present platforms are organized by independent unions, where organizing activities are limited to those platforms allowed by the companies. Since SPM platforms will probably be owned and operated by several oil and chemical companies, the firms are unlikely to turn it over to a single independent. Other unions such as the OCAW, and possibly the Operating Engineers—who have taken on interest in the exploration platforms—may attempt to organize the SPM platform employees. Their chances of success are probably low. The low number of employees, their physical location away from shore, a work schedule of seven days on and seven off, good personnel practices by parent companies in initial hiring—all work against the union in any organizational drive.

The dispatcher's position has no ready comparison to existing jobs, but the clerical/communicative nature of the job would probably result in reliable, corporate-oriented employees being placed in that position. The divers will be concerned with the underwater connections from buoy to shore. This duty could be performed by either permanent employees, or on a contractual basis to diving firms. The nature of the work and skill involved will probably result in diving wages, hours, and working conditions being regulated by the laws of supply and demand, rather than from the pressure of organized labor. The MEBA does represent a number of divers. The possibility of unionization on this front could have a significant impact beyond the terminal itself.

The crews of various mooring boats, launches and work boats are broadly grouped into the category of SPM related employees. Once again, their work is similar to what is already being supplied to exploration and production platforms. Since it can also be contracted out, union activity and manpower problems, as they affect the SPM system, can be expected to be minimal. Any problems will reflect existing conditions.

The key SPM related employees are licensed deck officers. Preliminary estimates for a system consisting of a platform and five SPMs indicate the need for four or five mooring masters and ten or eleven assistant mooring masters.

The mooring master's job is somewhat analogous to a docking pilot who assists the vessel's captain in approaching the SPM. The assistant mooring master's duties are almost those of an advisory chief mate. During the mooring he is stationed on the bow and assists in supervising the tying of the head lines to the SPM. Once it is moored the mooring master leaves the vessel while the assistant remains on board to look out for the terminal's interests and function as a watch officer. He sees that the vessel remains properly tied up, that hoses are connected properly, and that the oil flow to the platform is maintained. The difference between this officer and a relief mate common to shoreside oil docks is that he is not expected to supervise the vessel's crew in its pumping operations.

The position of these officers will be unique. They will be the only ones connected with the SPM system to have any direct and continuous interaction with the vessel's officers and crew. In addition, their training would have to consist of traditional shipboard experience as well as specialized training with the operation of SPMs. The manpower implications of this will be considered in a later section of this chapter, as they are tied to the vessel operations segment of the deepwater terminal system.

Tanker Registry and Some Initial
Labor Union Considerations

Tanker registry is an important consideration in examining the role of various unions in tanker operations as well as in operation of the terminals. Whenever a tanker makes a voyage between two domestic ports, such as Valdez, Alaska, and Long Beach, California, the Merchant Marine Act of 1920 requires that the tanker be of United States registry. On the long international runs that comprise the major portion of the emerging marine petroleum transportation system described in Chapter 2, the relative absence of large United States flag tankers means that foreign-registry tankers will be the major user, at least initially, of any deepwater ports off the East and Gulf coasts.[2]

Foreign-flag tankers could either be chartered or owned by the oil companies that form the parent company of the deepwater terminals. Because of the lower operating costs of such foreign-flag vessels (see Chapter 3), the oil companies would naturally prefer their use as the major carriers of imported oil. Not surprisingly, some independent tanker operators and United States maritime unions take the opposite view. The deepwater terminal interests were relieved to hear of the passage of a compromise version of the Deepwater Port Act which enables them to begin construction; but President Ford's pocket veto in January 1975 of the Energy Transportation Security Act, which would have eventually required that United States ships carry thirty percent of all imported oil, has forced the maritime unions to assess their strategy.[b] The potential combination of these acts, plus the Merchant Marine Act of 1970, which provides construction subsidies for bulk carriers, would have created a variety of vessels calling on the deepwater ports.

In addition, one cannot expect the United States maritime unions to be idle if their tanker employment continues to shrink while foreign jobs increase. At some point additional attempts will be made to either organize foreign seamen,

[b]U.S. Congress, House Committee on Commerce, *The Energy Transportation Security Act of 1974*, 93rd Congress, 2nd Session, N.R. 8193; hereinafter referred to as the Transportation Act, if signed into law would have required 20 percent of the gross tonnage of all oil imported on ocean vessels to be carried on flag tankers. The percentage would have increased to 25 percent in 1975 and 30 percent by 1977. U.S. Congress, Senate, *Deepwater Port Act of 1974*, 93rd Congress, 2nd Session, H.R. 10701, October 9, 1974. The Transportation Act was vetoed in January, 1975. The *Deepwater Port Act of 1974* was signed into law on January 4, 1975.

create closer alliances with foreign seamen unions, or harrass their operations somewhere along the delivery system. Since the courts are somewhat undecided on the rights to picket a foreign-flag vessel docked at a United States port, it is impossible to determine the rights of a "picket boat" around a SPM that may be forty miles from shore.[3]

Consideration should also be given to whether the union is independent or affiliated with a major union organization. Had the Energy Transportation Security Act been passed, oil companies would still have had the option of carrying their oil on United States flag vessels which some had planned to do. The significance here is that, with a few exceptions, most United States flag tankers owned by the major oil companies are organized by a variety of independent unions, as opposed to the traditional, AFL-CIO affiliated, maritime unions. Table 7-1 lists many of the independent labor associations; one can see that they are usually organized around individual tanker owners. Exxon, for example, has separate associations of licensed officers, radio officers, and tankermen.

This, in turn, creates another problem because of the possible rivalries and organizing efforts between independent unions and the affiliated maritime

Table 7-1
Independent Labor Associations[a]

Independent Association	Company
Jersey Standard Tanker Officers' Association Esso Radio Officers' Association Esso Tankermen's Association	Exxon
Texas Tanker Officers' Association Texas Radio Officers' Association	Texaco, Inc.
Mobil Tanker Officers' Association	Mobil Oil
Tidewater Tanker Officers' Association Tidewater Tankermen's Association	Tidewater Oil
Deepwater Officers' Association	Cities Service
Atlantic Maritime Officers' Association Atlantic Maritime Employees' Association	Atlantic Refining
Sun Marine Licensed Officers Association Sun Marine Employees' Association	Sun Oil
Sabine Independent Officers' Association Sabine Independent Seamen's Association	Sabine Transport.
American Tanker Officers' Association	American Trading & Production Co.

[a]Each of these associations negotiates directly with its respective employers.
Source: Maritime Administration

unions. Under the present system of operations seamen and officers belonging to various independent unions have very little contact, due to the fact that their vessels generally travel between company-controlled docks, where union organizers can be kept at a distance. This will not be the case with five or more vessels moored to a cluster of SPMs. Some interaction and comparisons are bound to take place during the discharging operations.

Maritime Union Rivalries

The affiliated maritime unions are notorious for their rivalries; and those rivalries are not subsiding. The marine petroleum transportation system is going to be affected by rivalries because they are moving away from the dry cargo part of the industry and toward the fleets owned by independent tanker operators. Should the major oil companies decide to let the independent operators carry the United States flag portion of oil imports, or in some cases sever their tanker fleets into separate corporate entities, the union rivalries could materially affect a significant proportion of deepwater terminal deliveries.

The history of the maritime union rivalries has been well documented.[4] The rivalries have centered mostly around ideologies, personalities, and shipowner associations. (It is not the purpose of this study to trace this history, but to point out the significant events that led to the present situation and its possible impact on the deepwater port system on the East and Gulf coasts.) The ideologies basically grew out of the American Federation of Labor (AFL) and the Congress of Industrial Organizations (CIO) affiliation with various unions for licensed officers and unlicensed seamen. These unions are arranged in Table 7-2. Basically, the Masters, Mates and Pilots (MMP), Seafarers International Union (SIU), and Radio Officers' Union (ROU) represent the AFL; the Marine Engineers Beneficial Association (MEBA), National Maritime Union (NMU), and American Radio Association (ARA) are on the CIO side. Since World War II the seamen's unions have been led and controlled by fairly powerful individuals. The alliances of the officers' unions with the unlicensed unions have often been based on their respect for or fear of these officials. Historically the MMP and MEBA were not jurisdictional rivals, and had a fairly cooperative relationship, in that their respective jurisdictions did not overlap, nor had they ever attempted to move into each other's territory. The officers' unions also had members serving on both NMU and SIU vessels. This made them an asset to the unlicensed unions, for in the event of a strike any licensed union could stop all vessels. On the other hand, the licensed unions needed the muscle of a larger unlicensed union to enforce their interests.[5]

In general, the SIU is the dominant offshore union on the West Coast, while its East Coast membership is concentrated on unsubsidized carriers. With the exception of the Cities Service fleet, its tanker contracts have been with

Table 7-2
AFL-CIO Affiliated Unions

I. Licensed Officers		Represents
MMP	International Organization of Masters, Mates and Pilots (Marine Division of International Longshoremen's Association)	Deck
MEBA	Marine Engineers Beneficial Association	Engine
AMO	Associated Maritime Officers (District 2, MEBA)	Deck
BMO	Brotherhood of Marine Officers (NMU)	Deck and engine
ARA	American Radio Association	Radio
ROU	Radio Officers' Union	Radio
II. Unlicensed Seamen		
NMU	National Maritime Union	
SIU	Seafarers International Union	
OCAW	Oil, Chemical and Atomic Workers [a]	

[a]Contract with Mobil Oil Company

independent or tramp operators. The membership strength of the NMU has been subsidized freighters, although it does have tanker contracts with some major oil companies and carriers such as Texaco, American Trading and Production Company, and Trinidad Tankers.

Because of the difference in the types of shipping companies the unlicensed unions serve, they have historically tended to differ in their views on collective bargaining. With a subsidized base, Joseph Curran, the NMU president, was able to concentrate on increasing wages and benefits for his members without being overly concerned about the demise of a few marginal operators. Paul Hall, the SIU president, has always had to be concerned with manning problems and the effect of any demands on the unsubsidized shipowner that could result in the loss of jobs for his membership. As a result, the SIU is often accused of signing "sweetheart contracts" (one that provides wages below the prevailing scale) to get a ship organized. Many owners disagree with such accusations, and contend that the SIU is more realistic in its approach.[6]

Union alliances will now be considered. Until 1956, unions generally aligned themselves along the AFL-CIO split. In 1956, the NMU agreed to a contract with the American Coal Shipping Company that would have them sailing on ships with members of the Brotherhood of Marine Officers (BMO). At that time the BMO was affiliated with District 50 of the United Mine workers. (District 50 was the catchall organization under which the United Mine Workers organized nonmining occupations that would ordinarily come under the jurisdiction of a rival AFL-CIO union.) This action was denounced by both the MMP and MEBA, and resulted in the MEBA being driven from their alliance with the NMU. The implication of that action has become very apparent as time passed. One problem for the MEBA was that, in 1950, the SIU had set up the AMO as an AFL rival of the MEBA. The AMO never made significant inroads into the MEBA's jurisdiction, but when the MEBA found itself alienated from the NMU it had no choice but to sue for peace with the SIU in return for some of the muscle it needed to fight the BMO. The end result was the merger of the AMO into Great Lakes District Number Two of the MEBA.

Structurally the MEBA is organized into districts; at the time of the merger District Two was one of the smallest. The merger gave the former SIU group an autonomous base within the MEBA. This has led to the MEBA and SIU becoming so tightly aligned that some people feel that the engineers have become subservient to the SIU.

The MMP remained within the SIU's sphere until 1961, when differences over negotiation strategies and personalities resulted in the MMP breaking from the SIU. This set of negotiations was quite complicated and involved a series of problems that are beyond the scope of this study. Its importance lies in the fact that it was the start of a series of events that are presently coming to a focal point that will affect operations of vessels serving deepwater terminals.

Prior to 1961 the only major employer group involved in East and Gulf Coast labor negotiations was the American Merchant Marine Institute (AMMI), which was composed primarily of subsidized operators and coastwise tanker companies. The unsubsidized companies that belonged to the AMMI did so mainly because of their NMU contracts. The tanker group included both small tanker companies and the marine divisions of the large oil companies. Unlike the dry cargo operators, they did not bargain as a group until the creation of the Tanker Service Committee (TSC) in 1962. The TSC was organizationally separate from the AMMI but remained closely allied with it. For all practical purposes the dry cargo and tanker negotiations have continued unchanged.[c] The unsubsidized cargo and tanker companies with SIU contracts generally had no formal employer association, until the 1961 creation of the American Maritime Association (AMA). Prior to this several of the Gulf Coast companies had a bargaining committee. The AMA caused little change other than to legitimize a situation where many small companies got together within what is acknowledged to be an SIU-dominated association, and, with few exceptions, accepted the union's demands.[7]

The issue in the 1961 negotiations centered around an SIU-MEBA-MMP push for expansion of jobs through the right to organize crews of United-States-

[c]In 1969, the AMMI merged with the Committee on American Steamship Lines and the Pacific American Steamship Association to form the American Institute of Merchant Shipping. Basically these groups contained the same company members from the subsidized sector of the industry. *New York Times*, February 28, 1969, p. 78.

owned foreign-flag ships, an employer agreement to lobby for a limitation on the amount of crude oil imported on foreign-flag tankers, and a push to modify the subsidy law to include bulk carriers. The NMU and its allies sought the traditional demand of wages and benefits. Had the MMP not left the SIU coalition, a situation could have been reached where all the SIU ships were sailing while the MMP kept all the NMU ships tied up on strike. The employers would have had very little to say in the matter.

Since 1961, the rivalries have continued. The SIU-MEBA coalition retaliated against the MMP through the admittance of deck officers in the AMO. The MMP and NMU remained cooperative, but the presence of the NMU-affiliated BMO made a close alliance impossible, particularly after the BMO started making advances toward organizing some MMP ships. Finally in 1971, the MMP found itself with no place to go among the offshore unions, and affiliated with the ILA.[8]

By 1974 the situation in the industry had reached a point where the rivalries could be solved. Some unions may be materially damaged by the solution; it may therefore be a subsiding of the battle rather than a real solution. The central problem is basically economic. The maritime unions are all relatively small, with a rather high dues structure. Manpower in the domestic industry is not growing; unions have had no material success in organizing outside of their traditional areas. Simple economy is increasingly dictating unity.

Of greater importance are the changes occurring in the industry itself. The laying up of all United States flag passenger vessels and reduction in the size of the subsidy fleet has materially hurt the NMU and its membership. On the other hand the SIU, which suffered the greatest loss of jobs due to the shrinkage of the United States flag fleet and the loss of jobs to foreign-flag vessels, is experiencing a renaissance of sorts, due to the upsurge in the construction of bulk carriers subsidized by the Merchant Marine Act of 1970. The NMU's weakening position, coupled with the retirement of NMU president Curran, have combined to make the 1970s an auspicious time for union alliance. A series of unity meetings has been taking place among the leaders of all the AFL-CIO-affiliated maritime unions. By late 1974, the state of the affiliated sector of the industry was highly fluid. Reports of progress in bringing the NMU and SIU closer together were accompanied by a report that the NMU-affiliated BMO was going to affiliate with the MMP.

In December, the MMP-BMO affiliation was ratified. The exact terms of the affiliation are not public but if the BMO comes under MMP control rather than as an independent entity under the ILA umbrella, it would give the MMP a source of engineering officers and increase their power position relative to the MEBA. As of early 1975, the effect of this affiliation upon the cautious optimism that surrounded the unity talks was unclear.

In addition to considering the history and current status of union rivalries and alliances, a great deal of consideration must be given to manpower training.

Supertankers calling at deepwater terminals are in a physical sense the "pipe" of marine petroleum transportation; they will in reality be no more effective than the crews that man them. One of the results of the series of negotiations that took place during the 1960s was a great interest on the part of the officers' unions to get directly involved in training and advancement. This was made possible by the creation of joint union-management funds negotiated into contracts. Both the MMP and MEBA have used these funds to create their own schools in Baltimore, where they train people for original licenses and for upgrading skills.[d]

Probably the most advanced school in the industry is the Maritime Institute of Technology and Graduate Studies (MITAGS), run by the MMP. The school has a heavy investment in a variety of simulation equipment. This places them in the advantageous position of having some of the latest training devices of this type, at a time when shipboard handling and safety has become a major issue, especially with respect to tankers.

In addition to considering union rivalries and training requirements, it would be remiss not to consider the political impact of maritime unions. Historically the maritime unions have had a far greater political impact than their numbers would indicate. During the last decade that impact has increased substantially. The prime mover in this has been the SIU; the MEBA is a close second. All maritime unions have been politically active; they certainly contribute to candidates who favor their points of view. Few unions can match the scale of the SIU and MEBA political operations and funds. The propriety of these funds and how they are amassed and distributed has been questioned.[9]

Political activity and spending on the part of the maritime unions continues to rise and is certainly an important political factor. The unions generally supported Deepwater Port legislation and were the primary backers of the Energy Transportation Security Act.[10] The Energy Transportation Security Act and the Merchant Marine Act of 1970 represent the 1961 goals of the SIU and AMA.

Maritime unions strongly favor deepwater terminals for added jobs as well as for possible national benefit.[11] Had the Energy Transportation Security Act and Deepwater Port legislation both been passed the unions would have achieved a major combination of legislative victories: deepwater terminals to import oil, and a requirement that a significant part of that oil be carried on United States vessels. This is in addition to the Merchant Marine Act of 1970, which provides a system of government subsidies for building the ships that will be required. The veto of the Energy Transportation Security Act has thwarted this goal but it has not eliminated it. Indeed, one can expect the issue to continuously arise until

dThe various schools set up by such funds are administered by joint union-management boards of trustees. Technically they are independent of the unions but their primary policies and direction appear to follow the desires of the union officials; therefore such schools will be referred to as "union" schools in this chapter.

such time as an equivalent bill or compromise measure has been reached. The public and private commitment to the goal of import quotas on foreign flag vessels makes retreat from this position close to impossible.

The Impact of the Maritime Unions
on Deepwater Terminal Systems

It is highly doubtful that the maritime unions will have a direct influence on the operations of a deepwater terminal. Their major operational impact may already have been made via support of the Deepwater Terminal legislation and the Energy Transportation Security Act. The primary impact of the maritime unions will probably be on the tanker delivery system to the terminal. In anticipation of the passage of the Energy Transportation Security Act, a variety of things occurred. In the unionized sector of the industry an immediate move on the part of the various unions to gain recognition on any new ships took place. Because such items as living spaces and conditions are an integral part of marine labor contracts, owners generally negotiate with and recognize a union during the vessel construction stage. With wages, working conditions, and benefits at a fairly high level, the emphasis of the unions has been on jobs for their members. It is difficult for union officials to negotiate reduced wages, but they have, and are willing to move on the less obvious issues of work rules and manning scales. The latter are of particular importance to the owners, and one can expect fewer men on supertankers than are on existing tankers. The SIU has traditionally been willing to follow this pattern in order to gain more jobs for its membership. Indeed, the SIU-MEBA-AMO contracts for new vessels have reflected this to such a degree that the MMP has had to waive those portions of many of its tanker contracts which call for a fourth mate.[12] The MMP fought long and hard to get the extra mate on the tankers, but now officials reluctantly admit that they must accept reduced manning or lose the vessels to the AMO.

Like the MMP, the NMU has not favored reduction of manning scales or the losing of previously gained work rules, but the erosion of their subsidy base has altered their position. Depending upon the progress of the unity meetings, the NMU can be expected to match any offer made by the SIU to a prospective shipowner. Had the BMO remained an NMU affiliate, the NMU would have been able to match the SIU's offer of a top-to-bottom contract. With the BMO leaving the NMU, the MMP is in a position to offer the tanker operators competitive single contracts for deck and engine officers. Should the scramble for jobs intensify, the MMP, without the BMO source of engineer officers, would find itself unable to obtain any new contracts, for the ILA could be of little help to them in organizing a tanker that never comes to port. The question then remains of what the NMU can or will do without a controlled source of officers. The industry has a way of proving most predictions to be wrong, but it would appear that the BMO spinoff is a negotiating fallback position for the NMU. Without the BMO, they are not a threat to the AMO, and their merger with the SIU is simplified. Should the merger talk sour, they could rebound to the ILA-MMP camp and create an equally powerful rival to the SIU and its allies.

Given these factors, it is quite apparent that the operators of United States flag tankers calling on deepwater terminals could benefit from maritime union rivalries in the short run. In spite of their acceptance of reduced manning, almost all union officials have voiced opposition to it on the grounds of individual fatigue and safety. Because of the impact of independent fleets they are reluctant to attempt to reverse the trend through collective bargaining as that would put the unionized portion of the tanker fleet at an economic disadvantage. Across the fleet manning increase can only come through regulatory agencies and this is where the pressure from labor will be directed. Long-run implications are harder to predict, but the increase of manpower costs should be as rapid as it has been in the past.

The reaction to the possible expansion of captive fleets should be considered next. As noted above, the Energy Transportation Security Act did not preclude the major oil companies from expanding their fleets of independently organized vessels[e] and several ships were on order prior to the veto. If the expansion is gradual and not immediately visualized as being at the expense of the unions, it could be beneficial to the oil companies. There is little apparent difference between the wages and working conditions on affiliated and independently organized tankers; however, the benefits of having an independent union go beyond wages. The combination of union ideology and questionable management attitudes has resulted in the officers and seamen of unionized vessels having very little corporate loyalty. Their pay increases, benefits, and job assignments come from the union; there is no reason for them to look to the companies for anything besides a temporary place of employment. This is not the case with the independent one-company unions, where employees tend to be more flexible in the interpretation of work rules and more willing to expend that immeasurable extra effort, if only because they have no place else to go for employment.

Both union organizing and the type of training personnel will receive should now be considered. Possible union organizing will be discussed first. The proximity of union and independent vessels around a cluster of SPMs will take place no matter which portion of the United States fleet expands. This will inevitably lead to closer comparisons of wages, working conditions, and above all, job security; it could erode the independent unions' control over their membership. If the corporate fleets greatly expand, a renewed interest in union organization can be expected. The MMP may lead this drive, if only as a defense

---

eFor example, Mobil Oil Corporation, which has an interest in both of the proposed Gulf Coast deepwater terminals, has already applied for a construction-differential subsidy to build a United States flag ULCC. Mobil Oil Corporation, *Annual Report*, 1973, p. 21.

against inroads of the AMO and BMO. At the same time, those major oil company fleets with NMU or SIU crews could find themselves under a great deal of pressure to have their officers affiliate with the BMO or AMO. The precedent for this type of action can be seen in several of the subsidized fleets where the masters were incorporated into the MMP contracts as a result of union pressure on the companies. In practice, these moves are already taking place at an accelerating pace. In late 1974, it was announced that the Exxon officers had decided to affiliate with the MMP. Although the members of the Exxon fleet rejected the affiliation by a narrow margin the prospect was a major break-through in this sector of the industry, and could signal the wholesale erosion of the independent tanker unions. Organizing drives have begun on other fleets, and some of the officials of the independent unions are rumored to be talking to the MEBA about affiliation. Organizing drives and inquiries such as this are not new; they have been a periodic occurrence for decades. Unlike the past, these are far more serious and involve more give and take on both sides. They cannot be ignored as they have been in the past.

The discussion will now turn to training. The capabilities of the personnel involved in manning the emerging marine petroleum transportation systems are obviously of critical importance to the environmental safety and economic success of the system. Their training is certainly an important factor. Any expansion of the oil company fleet is going to involve some manpower problems, especially in the training and development of officers. The traditional method of training seamen to officers via experience is coming under increased questioning due to the size and complexity of the vessels involved. As discussed earlier, newer training methods are necessary; simulators and various hands-on training devices appear to be the answer. Given the cost and environmental impact involved in tanker operations, it is quite probable that proven ability via simulation training will become part of the licensing process.

This is true for both the SPM mooring masters as well as the ships' officers. At present the mooring masters are trained on the job at foreign deepwater ports, but as the number of ports increase this will become a very costly and inefficient way to train people. Thus, the companies must find both a source of manpower and a way to train them in the specialized handling of ships designed to call on deepwater terminals.

The supply problem could be taken care of by the various maritime academy graduates who are experiencing difficulty entering the unions. The orientation and training of academy graduates cause them to look initially to the employer rather than to the union for direction and guidance; an attitude any reasonable employer would find desirable. The academies do not have the extensive modern training equipment considered here, although a simulator is being built at the United States Merchant Marine Academy. Union training schools do have some simulation equipment, and are in the process of planning for their own ship handling simulators.

The oil companies therefore have a problem. They are not likely to send their employees to a union school. Indeed a key attraction for the affiliation of Exxon's independent union with the MMP was the availability of training at MITAGS. The oil companies have some simulator training facilities in foreign countries. United States officers have been and are continuing to be trained there, but these facilities must serve the oil companies' foreign officers in order to divert the continuous charges that foreign flag vessels are poorly officered and manned. They cannot set up their own United States school for new officers without antagonizing the Federal Government and the various states which have maritime academies. The only short run answer for the oil companies is to hire academy graduates and increase their support of those schools. This is apparently what is happening; for it has been reported that one major oil company hired twenty-eight 1974 graduates of the United States Merchant Marine Academy.[13]

On the surface, company support of the academies is a fairly safe alternative. The maritime unions have traditionally mistrusted the academies which is one of the reasons they have created their own schools which can emphasize union philosophy along with the latest training methods. In addition, the academies have been a source of added manpower, even though shipping employment has declined. If the decline continues the unions will, as they have in the past, push for the elimination or curtailment of the academies. They are now in a position to justify this demand by noting the presence of their own nontax-supported schools. If the oil companies' support of the academies grows, the union can argue the issue of the federal and state subsidy of the manpower of the oil companies. Still, support of the academies is basically the only alternative open to the companies, for if the academies are curtailed the companies would have to go through the fairly expensive process of establishing their own United-States-based schools for a limited number of entrants into the profession. Thus, the growth of tanker fleets is going to result in a definite impact on the operation and philosophy of the maritime academies.

Upgrading and updating of licensed officers will continue to be an issue. Changes in training and proposed new programs are continuously being developed. It appears as if the companies are arranging themselves along a spectrum of alternatives to the training issue. Some continue to hold that on-the-job training is sufficient, others are supplementing that with short term seminars, while a few are making moves toward investing in some form of simulator training. Given the range of public, union, and private corporate training that is in existence or contemplated, it is quite possible that a situation could develop where a proliferation of limited simulation facilities will be developed in the near future.

Consider now the reaction to expanding the use of foreign-flag vessels. With or without the passage of the Energy Transportation Security Act, a significant number of foreign-flag vessels will be calling on deepwater terminals. Indeed, many people in the oil industry would prefer that they be the only vessels using the ports. If that extreme were to take place, the previous efforts of the unionized sector to organize foreign-flag vessels would probably be redoubled, along with their support of foreign unions.

The passage of the Energy Transportation Security Act would temper the union push against foreign-flag vessels due to the availability of United States Flag employment. Without this attraction, the pressure for jobs and employment might force the unions into an all-out battle to eliminate the independent unions, while seeking some personnel representation on foreign vessels. They would most probably exercise their political ability in a concentrated effort to control and further regulate the activities and income of the international oil companies.

Given the "back-to-the-wall" position the maritime unions are finding themselves in and their proclivity for fighting anyone and everyone, deepwater planners cannot ignore their presence. The Energy Transportation Security bill had been called the bill no one wants; it behooves deepwater terminal operators to consider the consequences of both passage and failure. Either way there will be some long-run costs involved that will reflect on the efficiency and economy of deepwater terminals. The immediate effects of the veto are apparent in the continued use of foreign flag ships, and the cancelling of plans for building new ships in the United States. The long-run effect of this action remains to be seen.

The Human Factor

The operational and manpower issues surrounding the emergence of deepwater terminals that have now been discussed cover the known qualities of the problem, in that they deal with numbers and placement of jobs, union recognition, skills training, and the important ebb and flow of political maneuvering by all sides. The human factor—so critical in the operation of supertankers and deepwater terminals—has not yet been discussed in any detail.[14]

Training throughout the marine petroleum transportation system involves engineering- and skill-oriented tasks. The problems the terminals—and in fact the entire system—face are not engineering ones, but human ones. These factors involve exchanges that take place along the interfaces of the system. The ability of a union to organize the workers in the system rises with the insensitivity of management to the social and behavioral needs of its employees. This is only one part of the problem, for the behavioral issues range well beyond unionization.

The environmental issue has been the major stumbling block to the passage of deepwater terminal legislation, and for the most part technology has answered these objections. Unfortunately, technology is not the major cause of oil spills, strandings, and collisions. All too often they are attributed to "human error." Unless the managers throughout the system are versed in the skills of management and communication, problems can occur at any of the transfer points.[15] Changing attitude of United States employees may bring about reaction to task-oriented leadership of the past.

The understanding of social groups and human relations will also become very important in operating deepwater terminal systems. The seven-day-on, seven-day-off pattern of the platform workers should not create any new problems that are different from those occurring on exploration platforms. The period of time the groups are together is not excessive in relation to the time they have off. The joining of these work crews with the licensed mooring officers could upset the existing patterns of platform life.

The tankers present a different problem. Long voyages are no longer as common to seamen and officers as they once were. For all practical purposes tankers calling on deepwater terminals may never see land. The long-term effect this will have on the crews is not really known. The crew of a tanker is a collection of humans who mold themselves into a small society over the period of a voyage. Changes in the manning scales will make these societies even smaller, and create greater stress. As the crew size declines, its members become more interdependent, while the pressure for individual conformity increases. A tanker crew of forty or fifty can isolate a deviant member without harming the mission of the vessel. A smaller crew can do the same, but the impact of the deviant among a crew of twenty can impair or hamper the safety and efficiency of the vessel.

Officers who have had to face psychological problems and intergroup conflicts on long voyages generally confess to being unprepared. They have been forced simply to react to such problems, as they did not know how to analyze or prevent them. The industry needs to recognize that human behavior problems are as common to the tanker as they are to the refinery and factory. If progress is to be made in the safe and efficient delivery of imported petroleum from the terminal system, then the industry will have to divert a minute portion of the millions of dollars these systems are going to cost to the effective education and training of system personnel in management and the behavioral sciences.

Such training can be found in every business school, but is virtually absent throughout the schools associated with the marine industry, including in many cases the maritime academies. This is in spite of the fact that the average officer coming out of any marine program has responsibility over more men, equipment, and total financial investment than the average graduate of even the best business school. If the emerging international system of supertankers and deepwater terminals has a weak link, this is probably it.

The Labor Perspective

An analogy can be drawn between the manpower issues surrounding deepwater terminals and the sea surrounding the actual terminal. On the surface all is calm and orderly; all of the real movement is taking place below the surface. The manpower needs of the overall system—the tanks and onshore refineries, the deepwater terminal system, and the supertankers that together form the "pipe" described in Chapter 1—can be discussed from the labor perspective. Skill training and necessary technical knowledge can be quickly secured from existing sectors of the oil industry. Beyond this, planning becomes more amorphous, particularly with respect to the tanker delivery system. The variety of unions involved and their complex and deep-seated rivalries could cause an infinite number of reactions to the operators of deepwater terminals. Looking to the past is of little help, as the unions in the industry have not been noted for their rationality. Reflecting on this fact one former secretary of labor said something to the effect that: "Labor relations are worse, collective bargaining works less well, and there is more violence and less reasoning in the maritime industry than in any other."

If deepwater terminals are to become the gateway joining the United States to the international marine petroleum transportation system, they are vital in overcoming the economic and physical barriers to entry that have existed with conventional United States ports. But deepwater terminals—functioning as the gateway to the "pipe"—may also serve as the focal point of a renewed merchant marine. This means that all interested in the planning and operations of deepwater terminals must be both knowledgeable of and sensitive to the collective bargaining environment. The unionized sector of the industry will want to be involved in deepwater ports, whether or not oil companies and terminal operators desire their involvement. A primary demand of unions will be for jobs. Failing that, increasingly adverse legislative activity is predictable. The vast political power and connections of the maritime unions simply cannot be ignored as the deepwater terminal system comes into being and is allowed to grow.

Training will remain an important factor in the system. This is especially true in the handling of tankers where the increased use of simulators can be predicted. Depending upon the vessel and union involved, this training can be expected to take place in corporate- or union-oriented schools, thereby creating tensions within the industry.

Finally, little is known about the effect of the closed working environment of large supertankers on the human behavior and the leadership of personnel who go to sea on these ships. Certainly this is so of Americans, as very few have sailed on such voyages. Although vast sums have been expended to measure maneuvering characteristics and conduct strain gauge studies, very little effort has been centered on managerial skills. At the simplest level, what might happen if a watch officer is afraid to call the master?

This is not to imply that human factors are more important than sophisticated technology. The opposite is true; in fact, human factors take on increasing importance because of the advances in technology that are probable in the system of supertankers and superships that is emerging. If the United States is to be a part of this system, and to obtain petroleum imports discussed in Chapter 1 this appears necessary, then human factors in the system will have to be considered in great detail. Management and behavioral science education and training for officers in the industry is just a first step in overcoming potential "human error." The labor aspects of deepwater terminal systems—tankers and terminals—are of extreme importance.

8 Meeting the Challenge

The Goal of Self-Sufficiency

The energy crisis and all of its resultant ramifications has caused the policy planners of the United States to consider the goal of becoming self-sufficient in energy. If this goal is attainable, the idea of building deepwater ports is an academic exercise. As noted previously, the option of zero imports involves some very high-level commitments and sacrifices on the part of the private and public sectors of the economy. A more pragmatic point of view, rather than the emotional anti-arab and public-relations view, forces the consideration of more than simple oil flows. When these considerations are added to the problem, the presence of a deepwater port starts to assume some importance.

The first consideration is that oil imports will have to continue at some level while the United States moves toward its goal. During this interim, the oil flow will continue with smaller ships or lightering of supertankers off the coast. Either way is both inefficient and environmentally unsound. The economic feasibility of building and operating a deepwater port only for this preself-sufficiency period is questionable. Therefore, additional factors must be considered.

The second consideration is that the economic cost of self-sufficiency runs counter to the public's demand for what it views as reasonably priced fuel. Consequently, rather than ask for the severe sacrifices on the part of the public, it is far safer politically for the government to rely upon mild restrictions, education, and voluntary compliance. However, such a policy involves a time-consuming process, and during the interim impacts will have to continue at a fairly substantial level.

The third consideration relates to international politics. Self-sufficiency by itself will simply result in the delaying of imports until such a time as the reserves in the United States are depleted. What self-sufficiency does allow is a bargaining position to keep international oil prices within some reasonable range. To do this the United States must continue to be a consumer nation which can adjust its level of imports to the world market. Moreover, since world trade flows in two directions, no country can politically afford to cut itself off from an oil-producing nation and expect to continue exporting technology and machinery to that nation.

The oil crisis of 1973-74 has shown that the United States and other oil-consuming nations are unable to control or depend upon a single source of supply. In addition, there is some question as to the willingness of the

129

oil-consuming countries to deal collectively with the producing nations. The need for oil and the competition to secure a source of supply dictates against any one consuming nation being able to control totally the supply of a producing nation. Thus, the 1974 announcement of a major discovery of oil in Mexico was quickly followed by that country's proclamation that it would supply oil to any consumer at prevailing world rates. This simply underscores the fact that overland pipelines do not necessarily mean a secure source of supply. Unlike the one-way pipeline, the supersystem discussed in the previous chapters provides any consuming nation possessing a deepwater port the total flexibility of a readily available tanker pipeline to any (willing) producing nation.

These economic realities on the international scene mean that the possession of a deepwater port allows a nation to compete more effectively for a supply of oil. While the lack of a deepwater port does not exclude a nation from the market, it limits that nation's ability to compete for lower transportation costs that allow higher bids for oil in the market place.

Given the factors of consumption, international trade and politics involved in the problem of supersystems, the issue quickly becomes dependent upon the potential level of oil imports that would use the system. As can be seen from Chapter 2, this is variable over a considerable range. Where in this range the actual level of imports falls depends upon the assumptions of the forecaster and the amount of faith one possesses in the ability of the United States to achieve its goal of self-sufficiency. As noted, some level of waterborne imports are required throughout the range. Should these projections hold true then the real question is not whether deepwater ports should be developed, it is their number and location in light of the world tanker market and environmental considerations.

Probable Location of Deepwater Ports

The tables and analysis developed in Chapter 2 clearly shows that the Petroleum Administration for Defense Districts (PAD) that will require waterborne oil imports are the East and Gulf coasts of the United States. Using three of the cases developed by the National Petroleum Council (NPC), a demand for waterborne imports is predicted in 1985 over a wide range—0.814, 5.951, or 10.724 million barrels per day (MBD).[1] Most likely estimates are about 6 MBD. (See Figure 1-2.) Each NPC case can justify the building of a deepwater port, the first for political and competitive reasons in the world marketplace and the others out of the need to handle the volume of imports. Since those deepwater ports which are already being proposed call for a capacity of 3 MBD each, it is quite possible that the United States could find itself with excess importing capacity at the wrong locations.[2]

Location is a bigger problem. The East Coast, New England in particular, has been an energy importing area ever since coal replaced the waterwheel. No major refineries are located in New England, and proposals to create supersystems involving the construction of refineries and deepwater terminals have generally been opposed.

Refinery capacity exists in the rest of the PAD I Atlantic Coast region. A good case can be made for the location of a deepwater port serving the major refineries of the area.[3] Still, as was shown in the earlier chapters, a strong argument has been made to keep out deepwater ports on the grounds of potential damage to the area's recreational industry.

The situation on the Gulf Coast is nearly the opposite of that on the East Coast. Two consortia of oil, chemical, and pipeline companies have already been established, have overcome some initial environmental opposition, and as of early 1975, eagerly awaited the necessary permits to begin construction.[4] Offshore wells have been operating for a number of years in this area and the vast concentration of oil and chemical companies makes the development of deepwater ports an economic necessity, if one accepts NPC CASE III. Assuming a lower level of imports leads to the questioning of building two deepwater ports in the same district when the greater need for imports lies elsewhere. The answer seems to lie in the range of public acceptability rather than necessity or practicality.

Technological Aspects

Since there are more than one hundred deepwater ports in existence today, technology cannot be considered a constraint to superport activity. At present, it is not possible to detect small leaks from submerged pipelines. In addition, more work needs to be done on containment and cleanup of oil spills.

While technological considerations should not be a primary constraint on the review of environmental impact statements, lack of scientific knowledge may present problems in this area. More research needs to be done in evaluating the impact of oil spills on marine life. The problem of oil pollution has not been studied long enough to be able to analyze its effects over a several-year period.

Supertankers—Causes for Concern

The East and Gulf coasts are the key import areas which can only be served by a tanker system. The combination of supertankers connecting to a deepwater port system creates an extremely efficient and flexible oil delivery system. The earlier chapters have shown how the development of a deepwater port system can materially reduce the number of vessels needed and the costs involved in transporting oil. There are some very real environmental considerations involved in the continued operation of these tankers.

These factors must be weighed in light of the world tanker market. Since tankers come very close to operating in a perfectly competitive market, the economic impact of regulation becomes a major consideration. The economic savings are the greatest for ships that are the most inexpensively built and least regulated. Under a competitive system, the country that encourages this will find itself with the greatest number of vessels.[5]

In a competitive international market any environmental regulations dealing with tanker-generated pollution on the international waters they transit require international agreement for ship construction and operation standards. It is the very essence of the tanker trade that they transit waters of different nations at each leg of a voyage. Clearly, nations ratifying the new IMCO convention must enforce its rules on ships of nations that do not, so that tanker-pollution-prevention regulations do not allow "comparative advantage" to some competitors. Such an advantage on the part of less regulated carriers could allow the market mechanism to work toward lowering the environmental quality for all countries in the long run.[a]

The Economic-Environmental
No-Win Situation

There are a multitude of factors that surround deepwater port systems. If viewed solely as an economic problem, then one can choose a level of waterborne imports and justify a certain number of ports. Lower the level far enough and it can still be justified for economic or political reasons.

If the port can be justified then the tankers serving it can be as well on the basis of the enormous efficiency and savings generated. In actual practice it was the savings generated by the quantum growth in tanker size that created the need for deepwater ports in the first place. The tanker is the most inexpensive way to transport oil, and the deepwater port system makes it even more efficient.

At the same time, the presence of a deepwater port will open the United States up to an environmental risk that could be of catastrophic proportions. Should a supertanker serving a deepwater port have a collision, grounding, or some natural or manmade disaster off the shore of the United States, the environmental damage to the oceans and seashore could be disastrous. The environmentalists have a valid issue in questioning the construction of deepwater ports that would allow a potential hazard near our shores.

The answer to the extreme desire of some environmentalists for no deepwater ports is that the hazard is already present in several forms. The supertankers are

[a]For further information see Louis K. Bragaw, "Environmental Policy Formulation in Competitive Tanker Markets," *Decision Sciences Northeast Proceedings*, Vol. 3 (1974), pp. 31-34.

already offshore; there are environmental dangers inherent in the present delivery system as well as in the proposed no-deepwater port alternatives. In the absence of a deepwater terminal, some oil companies are delivering oil off the shores of the United States on ships that are in excess of 80,000 tons and cannot enter existing harbors. Their cargo is being lightered out at sea into smaller tankers involving all of the dangers present in the close maneuvering of ships and the probability of spillage during transfer. In addition, it is more expensive.

The present system of smaller tankers is also expensive and environmentally hazardous. Instead of one large spill, the rivers and harbors are subjected to innumerable small spills and accidents that collectively produce a great deal of environmental damage. Without a deepwater terminal, any increase in the level of imports means a greater number of vessels calling on United States ports and a corresponding increase in the possibility of accidents and spills. Note that these spills may occur closer to productive estuaries than is possible with an offshore terminal.

One alternative to a United States deepwater port is the building of a port in a neighboring country or in the Caribbean and transferring the oil by barge to the United States. This plan has the distinction of being unsatisfactory both economically and environmentally. The system would be more expensive to operate than a deepwater port and its efficiency would be questionable. It could also be an environmental nightmare. The risk of strandings and collision would be high due to the awkwardness of the numbers and types of barges involved. The increased number of transfers required would also mean a corresponding increase in the number of potential spills. In addition, there are significant political disadvantages in depending upon another country for a transshipment terminal.

The result is the fact that a purely environmental or economic viewpoint is unacceptable. Neither side can afford to win without passing on some very high potential hazards. The situation really calls for some reasonable compromises on both sides of the issue.

Necessary Compromises

Deepwater port systems can yield substantial savings even under the worst case examples given in Chapter 3. With the addition of unanticipated environmental safeguards, the system could still pay for itself if a sufficient volume of imports is maintained. Environmental savings are possible even below breakeven conditions, due to the reduction of the number of smaller tankers calling on United States ports.

The type of deepwater terminal to be used seems to have been decided in that both the Seadock system in Texas and the LOOP system in Louisiana call for the building of single-point-buoy (SPM) systems. This design has advantages over the alternatives of artificial islands and deeper channels.

With more than one hundred single point moorings in existence, this design has proven itself both economically and environmentally. The SPM can yield the economies necessary to make the system work. Of the various possible designs, the SPM requires the lowest initial cost, although its maintenance costs are higher than some other systems. The large amount of world-wide experience with SPM systems should help to facilitate the environmental review procedures required by the Deepwater Port Act of 1974.[b]

The SPM system is apparently the least damaging design from an environmental point of view. Construction of the SPM causes minimal disturbances and has no long-term impact upon the ocean bottom. As with all designs, the skill of the personnel operating the vessels and the cargo handling equipment is of paramount importance to environmental safety.

Supertanker Considerations and Compromises

The entire deepwater port question revolves around the ships using it. They are the most expensive part of the system, yield the great economies of scale that make it work, and are potentially the most environmentally damaging part of the system. The specter of a *Torrey Canyon* disaster off the shore of the United States must be dealt with realistically and without equivocation if acceptable deepwater port systems are to be developed. Risks are an inherent factor in vessel operations; the goal must be one of minimizing, not eliminating, risk. The ships must be considered in three respects; construction, operation, and size.

A well-designed, constructed, and tested ship is one of the surest ways of minimizing some of the major risks in tanker operations. Serious attention must be given to an entire series of considerations ranging from reserve power systems so that the ships can safely carry out their functions with minimal environmental damage, to on-board sewage treatment so that the sea around the terminal is kept free of pollutants.

The operations of the tankers must also be regulated. This involves going beyond the use of traffic control to minimize the possibility of collisions, to the personnel on the ships. Human error is too great a risk to allow the least regulated and trained officers and crews to set the safety standards on supertankers. In this area, the goal of minimum risk dictates that stringent training and proven ability be required of those persons who both sail the supertankers calling on deepwater ports and who are involved in the cargo operations of that vessel.

Finally, the combination of insurance rates, return on investments, and financial responsibility for spills may well dictate that an optimal range of

[b]See the Deepwater Port Act of 1974, Public Law 93-627, 88 Stat. 2126.

tanker size may exist for serving the deepwater ports of the United States.[c] Additional studies need to be made in this area of risk analysis, for there is a point at which the rate of return for a mammoth ship simply is not worth the environmental hazard that vessel represents, no matter how remote the probability of its having an accident.

There is no question that strict regulation will produce ships that cost more to build and operate. That fact will have to be accepted on the grounds that a sufficient volume of cargo and the efficiency of the system can meet the increased cost. If it cannot do so, then the country must be willing to bear its share of increased costs.

Labor and Manpower

Another key issue in the manning of supertankers and deepwater terminals is whether supersystem personnel will be unionized or non-unionized. Traditionally, almost all imported petroleum has been transported in foreign-flag tankers; thereby, personnel have been out of jurisdiction of United States labor unions. Given the political realities of the unionized sector, major efforts will be made to have supertankers built in the United States employ union personnel in serving the supersystem. Unions will continue to advocate cargo preference arrangements to increase the number of jobs for their members.

A Final Word

This Nation presently faces the possibility of a long-term energy shortage unless steps are taken to conserve the energy available, develop new sources of energy, and attain self-sufficiency in energy. These measures, however, will take time. And even when the capability for energy self-sufficiency is attained, it can be anticipated that we will be importing significant amounts of crude oil and petroleum products. If the crude oil and petroleum products are to be imported efficiently and economically, it is necessary that deepwater port facilities be constructed which can accommodate the new very large cargo carriers.[6]

This opening statement on a report accompanying the Deepwater Port Act to the House of Representatives sums up the situation of the United States in 1975. *The need for deepwater supersystems has been recognized.* Planning and development of deepwater supersystems must be realistic, timely, and responsive to national needs. This is the challenge of deepwater terminals.

---

[c]A recent study of oil spill statistics could find no such optimal sizes per se. See Louis K. Bragaw and Robert N. Stearns, "Environmental Costs and Optimal Tanker Size," paper presented at XXI TIMS International Meeting, San Juan, Puerto Rico, October 18, 1974.

National needs require that a deepwater port or ports be built; economic considerations must balance the environmental safeguards for the coastlines. The compromises that are required in designing and operating supersystems to allow for environmental quality are not beyond the pale of sound economics. What is really involved is sharing of savings generated in the supersystem. The potential size of environmental damage is so large that policy planners cannot rely upon the corporate goodwill of the users of the system. The avarice or negligence of a single minor operator could undo the collective responsible actions of the whole; therefore, stringent environmental safeguards are mandatory. Only by planning realistically, in terms of both economy and ecology can the United States properly meet the *challenge of deepwater terminals.*

Notes

Notes

Chapter 1
The Demand for Petroleum Imports to
the United States and the Challenge
of Deepwater Terminals

1. Energy Policy Project of the Ford Foundation, "Exploring Energy Choices: A Preliminary Report" (Washington, D.C., 1974), pp. 1-41.

2. U.S. Department of the Interior, "An Analysis of the Economic and Security Aspects of the Trans-Alaska Pipeline, Volume II" (Washington, D.C., 1971), p. 4-i-7.

3. Committee on U.S. Energy Outlook of the National Petroleum Council, "U.S. Energy Outlook" (Washington, D.C., 1972), p. 35.

4. Energy Policy Project of the Ford Foundation, p. 74.

5. Committee on U.S. Energy Outlook of the National Petroleum Council, p. 4.

6. U.S. Department of the Interior, "An Analysis of the Economic and Security Aspects of the Trans-Alaska Pipeline, Volume III" (Washington, D.C., 1971), pp. IV-14-26. See also R.R. Berg, J.C. Calhoun, and R.L. Whiting, "Prognosis for Expanded U.S. Production of Crude Oil," *Science*, Vol. 184, No. 4134 (19 April 1974), pp. 331-36.

7. M. King Hubbert, "The Energy Resources of the Earth," *Scientific American*, Vol. 224, No. 3 (September 1971).

8. Energy Policy Project of the Ford Foundation, p. 1.

9. Committee on U.S. Energy Outlook of the National Petroleum Council, "U.S. Energy Outlook."

10. Energy Policy Project of the Ford Foundation, p. ii.

11. Ibid., p. iii.

12. The National Petroleum Council, "Guide to the National Petroleum Council Report on United States Energy Outlook" (Washington, D.C., 1972), p. 6.

13. Energy Policy Project of the Ford Foundation, pp. 39-54.

14. Committee on U.S. Energy Outlook of the National Petroleum Council, pp. 17, 18.

15. Energy Policy Project of the Ford Foundation.

16. Kenneth Boulding, "The Economics of Spaceship Earth," in Henry Jarrett, ed., *Environmental Quality in a Growing Economy* (Baltimore, Md.: Johns Hopkins University Press, 1966).

17. Energy Policy Project of the Ford Foundation.

18. William P. Tavoulareas, "Growth vs. Stagnation" (New York: Mobil Oil Corporation), p. 16.

19. "NPC Urges Storage of 500 bbl," *Oil and Gas Journal*, Vol. 72 (September 16, 1974), pp. 34-35.

20. Zenon S. Zannetos, *The Theory of Oil Tankship Rates* (Cambridge, Mass.: The M.I.T. Press, 1966), pp. 174-85.

21. Ibid., see also Morris A. Adelman, *The World Petroleum Market* (Baltimore, Md.: Johns Hopkins University Press, 1972), pp. 102-6.

22. Adelman, p. 106.

23. Zannetos, pp. 142-43.

24. Christopher Hayman, "What to Do With All Those Tankers," *The New York Sunday Times*, April 14, 1974, p. F-1; see also James L. Rowe, Jr., "Administration Weighs Tanker Industry Help," *The Washington Sunday Post*, May 11, 1975, p. L1, 4.

25. S.G. Sturmey, *On the Price of Tramp Ship Freight Service* (Bergen, Norway: Institute for Shipping Research, 1965), see also Adelman, Zannetos and Rowe.

26. Lewis Beman, "Betting $20 Billion in the Tanker Game," *Fortune*, Vol. LXXX, No. 20 (August 1974), pp. 145-48.

27. Harry Benford, "Measures of Merit in Ship Design," *Marine Technology*, Vol. 7, No. 4. (October 1970), pp. 465-76; see also Zannetos, p. 127f.

28. Adelman, p. 130f.

29. Henry S. Marcus, "The U.S. Superport Controversy," *Technology Review*, Vol. 75, No. 5 (March/April 1973), pp. 49-57.

30. Fearnley and Egers, *World Bulk Fleet*, and *Large Tankers* (Oslo, Norway: Fearnley and Egers, 1974).

31. Ibid.

32. Hayman.

33. Louis K. Bragaw, "Environmental Policy Formulation in Competitive Tanker Markets," *Decision Sciences Northeast Proceedings*, Vol. 3 (1974), pp. 31-34.

34. Louis K. Bragaw and Robert N. Stearns, "Environmental Costs and Optimal Tanker Size," paper presented at the XXI TIMS International Meeting, San Juan, Puerto Rico, October 18, 1974.

Chapter 2
The Location of Deepwater Terminals

1. Henry S. Marcus, "The U.S. Superport Controversy," *Technology Review*, Vol. 75, No. 5 (March/April 1973), pp. 49-57.

2. Ibid.

3. Joseph D. Porricelli and Virgil F. Keith, "Tankers and the U.S. Energy Situation—An Economic and Environmental Analysis," ASME Publication 73-ICI-106 (New York: The American Society of Mechanical Engineers, 1973).

4. "Guide to the National Petroleum Council Report on United States Energy Outlook" (Presentation made to the National Petroleum Council, Washington, D.C., December 11, 1972), p. 13.

5. Ibid., p. 20.

6. U.S. Department of Transportation, *Refinery–Deepwater Port Location Study* (Washington, D.C.: U.S. Department of Transportation, 1974).

Chapter 3
Economic Aspects of Deepwater
Terminal Systems

1. Raytheon Company, *Massport Marine Deepwater Terminal Study, Interim Report Phase IIA* (May 1974), p. 80.

2. Ibid.

3. Offshore Oil Task Group, Massachusetts Institute of Technology, *The Georges Bank Petroleum Study*, Vol. 1 (Cambridge, Mass.: Sea Grant Project Office, 1973), pp. 46 and 51.

4. Peter M. Kimon, Ronald K. Kiss, and Joseph D. Porricelli, *Segregated Ballast VLCCs, An Economic and Pollution Abatement Analysis*, Society of Naval Architects and Marine Engineers, Chesapeake Section, January 11, 1973, and Part 2, *Segregated Ballast Aboard Product Tankers and Smaller Crude Carriers,* Department of Transportation, Coast Guard, February 1973. Note that in published form, Part 2 does not bear the names of any specific authors.

Chapter 4
The Environmental Perspective

1. *Deepwater Terminal in Delaware Bay*, Proceedings of Public Meeting, Department of the Army, Philadelphia District, Corps of Engineers, Philadelphia, Pa.: (March 31-April 1, 1970), pp. 31-33.

2. Ibid, p. 32.

3. *Tools For Coastal Zone Management*, Proceedings of the Conference (Washington, D.C.: Marine Technology Society, February 14-15, 1972), pp. x-xvi.

4. Ibid., p. xi.

5. Ibid., pp. xii-xiii.

6. *Marine Coastal Resources Renewal* (Augusta, Me.: State Planning Office Executive Department, July 1972), pp. 73-74.

7. "Port Growth Policies Abroad," *Water Spectrum* (Army Corps of Engineers, Winter 1971-72), pp. 2-3.

8. *The Environmental and Ecological Aspects of Deepwater Ports*, Vol. IV

(Department of the Army, Corps of Engineers, U.S. Deepwater Port Study, August 1972), pp. 26-28.

9. See Paul Eurlich "Eco Catastrophe!" *Ramparts* (reprinted by Boston Area Ecology Action, 925 Massachusetts Ave., Cambridge, Mass.), pp. 24-28.

10. *Environmental and Ecological Aspects of Deepwater Ports*, p. 27.

11. See *Preliminary Study of Six Mile Superport for Delaware Valley Council* (Hudson Engineers—Professional Engineers and Consultants, 121 South Broad St., Philadelphia, Pa. 19107, September 1972).

12. *Environmental and Ecological Aspects of Deepwater Ports*, p. 49.

13. United States Coast Guard, *Polluting Spills in United States Waters*, 1970, (Washington, D.C.: United States Coast Guard, September 1971).

14. United States Department of the Interior, *Minerals, Facts and Problems, 1970* (Washington, D.C.: Bureau of Mines Bulletin 650, 1970).

15. *Environmental and Ecological Aspects of Deepwater Ports*, p. 57.

16. Ibid., p. 58.

17. Ibid., p. 56.

18. J.S. Smith, *Torrey Canyon Pollution and Marine Life* (London: Cambridge University Press, 1968), p. 14.

19. Stanley E. Degler, *Oil Pollution, Problems and Policies* (Washington, D.C.: BNA's Environmental Management Series, Bureau of National Affairs, Inc., 1969), p. 11.

20. Ibid., pp. 2-20.

21. *Deepwater Terminal in Delaware Bay*, pp. 46-50.

22. *Deep Draft Vessel Port Capability on the U.S. North Atlantic Coast,* (U.S. Department of Commerce, Maritime Administration, 1970), p. 15.

23. *Environmental and Ecological Aspects of Deepwater Ports*, p. 20.

24. F.A. Smith, et al. *Fourteen Selected Marine Resources Problems of Long Island, New York: Descriptive Evaluations.* (Hartford, Conn.: The Travelers Research Corp., January 1970).

25. Degler, p. 2.

26. *Environmental and Ecological Aspects of Deepwater Ports*, p. 21.

27. Peter Hepple, ed., *Water Pollution by Oil* (Proceedings of a seminar held at Aviermore, Invernesshire, Scotland, 4-8 May 1970, London, Elsevier Publishing Co. 1971).

28. Louis K. Bragaw and Leo Jordan, "Coast Guard Offshore Oil Pollution Control Systems," *Offshore Technology Proceedings*, Vol. I (May 1974), pp. 549-58; see also *Environmental and Ecological Aspects of Deepwater Ports*, p. 45.

29. Ibid., p. 46.

30. Smith, p. 176.

31. Dillingham Corporation, *Systems Study of Oil Cleanup Procedures* (A Report to the Committee on Air and Water Conservation of the American Petroleum Institute, February 1970).

32. *A Deepwater Port Analysis, Delaware River Bay*, (Delaware River Port Authority, World Trade Division, Applied Research Bureau, Philadelphia, Pa., 30 May 1972).

33. Ibid.

34. Smith, p. 14.

35. *Environmental and Ecological Aspects of Deepwater Ports*, p. 43.

36. Smith.

37. Max Blumer, Testimony Before the Antitrust and Monopoly Subcommittee of the U.S. Senate, August 1970.

38. Lawrence G. Jones, et al. "Just How Serious Was the Santa Barbara Oil Spill?" *Ocean Industry* (June 1969), pp. 53-56.

39. H.D. Hasler, in E. Sondheimer and J.B. Simeone, eds., *Chemical Ecology* (New York: Academic Press, 1970).

40. J.H. Todd, et al., "An Introduction to Environmental Ethology" (WHDI-72-43, unpublished manuscript, Woods Hole Oceanographic Institute, Woods Hole, Mass., 1972).

41. K.J. Whittle and Max Blumer, in D.W. Hood, ed., *Organic Matter in Natural Waters* (Institute of Marine Science, University of Alaska, 1970).

42. Ibid.

43. Steven F. Moore, Robert L. Dwyer, and Arthur M. Katz, *A Preliminary Assessment of the Environmental Vulnerability of Machias Bay, Maine to Oil Supertankers* (Report No. MITSG 73-6 Massachusetts Institute of Technology, Cambridge, Mass. 02139, January 15, 1973), pp. 91-96.

44. Blumer, Testimony before the Special Commission for Studying Marine Boundaries and Resources of the Commonwealth of Massachusetts, February 4, 1972, p. 289.

45. Blumer, Testimony before the Antitrust and Monopoly Subcommittee.

46. Blumer, Testimony before the Special Commission, p. 290.

Chapter 5
Issues, Actors, and Onshore Impacts

1. *Deep Draft Vessel Port Capability on the U.S. North Atlantic Coast* (U.S. Department of Commerce, Maritime Administration, 1970), p. 7.

2. Ibid., p. 122.

3. Henry S. Marcus, "The U.S. Superport Controversy," *Technology Review* Vol. 75, No. 5 (March/April 1973), p. 55.

4. Ibid.

5. *Deepwater Terminal in Delaware Bay*, Proceedings of Public Meeting, Department of the Army, Philadelphia District Corps of Engineers, Philadelphia, Pa. (March 31-April 1, 1970), pp. 31-33.

6. Hugo Grotius, *Mare Liberum* (Margoffran trans. 1609).

7. *Submerged Lands Act*, 10 U.S.C. §§7421-26, 7428-38, 43 U.S.C. §§1301-03, 1311-15.

8. *Convention on the Territorial Sea and the Contiguous Zone*, Art. 2 completed April 29, 1958, 15 U.S.T. 1606, TIAS No. 5639, 516 U.N.T.S. 211 (effective September 1964).

9. Presidential Proclamation No. 2667, 10 Federal Reporter 12303, September 28, 1945. 59 Stat. 885-886.

10. North Sea Continental Shelf Cases, ICJ 3, 33 41 ILR 29 (1969).

11. United Nations Conference on the Law of the Sea; Convention on the Continental Shelf, Geneva (opened for signature April 29, 1958) [1964] 15 U.S.T. 473, TIAS No. 5578, 499 U.N.T.S. 311.

12. W. Friedman, *The Future of the Oceans* (1971), p. 33.

13. 33 U.S.C. Stat. 40 et seq. (1970).

14. 83 Stat. 852, 42 U.S.C.A., pp. 4321-47 (Supp. 1970).

15. Ibid.

16. P.L. 90-23, H.R. 5357, 90th Congress, June 5, 1967.

17. P.L. 92-574, 92nd Congress, October 18, 1972.

18. P.L. 91-604, 84 Stat. 1676; 42 U.S. Code Ann. 1857, 49 U.S. Code Ann. 1421, 1430; see also the National Ambient Air Quality Standards of 7 April 1971, 36 F.R. 1503.

19. *Coastal Zone Management Act of 1972*, P.L. 92-583 (The Bureau of National Affairs, Inc., 1972).

20. *Panama Railway Co. vs. Napier Shipping Co.*, 166 U.S. 280 (1897).

21. United States Constitution, Article III, Section 2.

22. 46 U.S.C. 740 (1958).

23. 46 U.S.C. 181 et seq. (1958).

24. 46 U.S.C. 183(a) (1958).

25. See *California vs. S.S. Bournemouth*, 307 Fed. Supp. 922 (C.D. California 1969).

26. P.L. 92-500, 92nd Congress, 1972.

27. 33 U.S.C. 407 (1964).

28. 83 Stat. 852, 42 U.S.C.A., pp. 4321-4347 (Supp. 1970); P.L. 90-23, H.R. 5357, 90th Congress, June 5, 1967; Commonwealth of Massachusetts Act 768, 1965, Act 228, 1965, and Act 44, 1968; Carl H. Bradley, Jr., and John M. Armstrong, *A Description and Analysis of Coastal Zone and Shoreland Management Programs in the United States* (University of Michigan, Sea Grant Program, Technical Report 20, 1972), pp. 39-50; *Coastal Zone Management Act of 1972*, P.L. 92-583, (The Bureau of National Affairs, Inc., 1972), respectively.

29. 33 U.S.C. 407 (1970).

30. 33 U.S.C. 407 (1970).

31. 33 U.S.C. 1161 (1970).

32. 384 U.S. 224 (1966).

33. 384 U.S. 226 (1966).

34. 362 U.S. 482 (1960).

35. 33 U.S.C. 411 (1970).

36. Florida Statutes Annotated, Section 376 (1972-73); Massachusetts General Laws Annotated, Chapter 130, Section 23 (1972); Delaware Code Annotated, Title 7, Sections 7001-7214 (1971), respectively from Leonard A. Lipson, "Offshore Terminals and Pollution of the Seas by Oil," in L. Lipson, L. Kaster, T. Porter, P. Carre, and G. Yeannaki, *Regulation and Jurisdiction in Offshore Port Development* (MIT Commodity Transportation and Economic Development Laboratory, Report MITCTL 73-8, June 1973).

37. Lipson, p. 27.

38. Ibid.

39. 15 U.S.C. 1 (1964).

40. 326 U.S. 1 (1945).

41. Thomas Porter, "The Effect of the Antitrust Laws on a Joint Venture to Finance, Build and Operate an Off-Shore Oil Terminal," in Lipson, et al., p. 72.

42. 211 U.S. 82 (1911); 221 U.S. 106 (1911); 224 U.S. 383 (1912) respectively.

43. 224 U.S. 405 (1912).

44. Nathan Associates, Inc., *U.S. Deepwater Port Study*, Vol. 1, Summary and Conclusions, August 1972, p. 34.

45. 15 U.S.C. 18 (1964).

46. Austin Stickells, *Federal Control of Business: Antitrust Laws*, (Rochester: The Lawyers Co-operative Publishing Co., 1972), p. 31.

47. Thomas Porter, "The Effect of the Antitrust Laws on a Joint Venture to Finance, Build and Operate an Off-Shore Oil Terminal," in Lipson, et al.

48. 83 Stat. 852, 42 U.S.C.A., pp. 4321-47 (Supp. 1970).

49. Sa H.R. 6750, 91st Congress, 1st session (1967) and S1075, 92nd Congress, 1st session (1969).

50. Ibid.

51. Frank P. Grad, *Environmental Law* (New York: Matthew Bender, 1971), pp. 13-1-13-11.

52. Ibid., pp. 13-2-13-3.

53. "NEPA—Reform in Government Decision-Making," *3rd Annual Report of Council on Environmental Quality*, August 1972, p. 223; see also Congressional Record 26568-91.

54. Ibid., p. 223.

55. *Freedom of Information Act*, P.L. 90-23, 1967.

56. Albert J. Rosenthal, "The Federal Power to Protect the Environment: Available Devices to Compel or Induce Desired Conduct," 45 Southern California Law Review 2 (May 1972), pp. 397-449; Joseph Sax, "The Search for Environmental Quality: The Role of the Courts," *The Environmental Crisis* (New Haven, Conn.: Yale University, 1970); "The Law and the Environment," *2nd Annual Report of Council on Environmental Quality* (Washington, D.C.:

Council on Environmental Quality, August 1971); "Some Do it Yourself Suggestions on How To Become an Effective Lobbyist," Office of Model Cities, Massachusetts Department of Community Affairs (mimeographed, n.d.); Frank P. Grad, "The Need For Integrated Approaches to Environmental Management" and "Environmental Litigation—The Effort of Private Litigants to Vindicate Public Rights," *Environmental Law* (New York: Matthew Bender, 1971); Janet Spencer, "Defending the Future: Sketches of Four Environmental Law Groups," *Juris Doctor* (February 1972); Irving Like, "Multi-Media Confrontation—The Environmentalists' Strategy for a No-Win Agency Proceeding," 1 *Ecology Law Quarterly* 495 (1971); B. Wiggins, "Applying the Fairness Doctrine to Environmental Issues," 12 *Natural Resources Journal* 1 (January 1972); "Corporate Advertising and the Environment," *Report of Council on Economic Priorities*, vol. 2, No. 3 (September 1971), respectively.

57. Sax.

58. Bostwick H. Ketchum, *The Water's Edge, Critical Problems of the Coastal Zone* (Cambridge, Mass.: MIT Press, 1972), p. 10.

59. *Regional and National Demands on the Maine Coastal Zone* (Boston, Mass.: New England River Basins Commission, 1971), p. 23.

60. Ibid.

61. Ibid.

62. Ketchum, p. 11.

63. *Deepwater Terminal in Delaware Bay*, p. 32.

64. Ketchum, p. 104.

65. Ibid., p. 106.

66. Bertrand L. de Frondeville, *Foreign Deep Water Port Developments. A Selective Overview of Economics, Engineering, and Environmental Factors.* Vol. I, II, III (Cambridge, Mass.: A.D. Little, Inc., December 1971), passim.

67. Ibid.

68. *Deepwater Terminal in Delaware Bay*, pp. 32-33.

Chapter 6
The Role of the Federal Government

1. Bertrand de Frondeville, "Port Growth Policies Abroad," *Water Spectrum* (Winter 1971-72), p. 5.

2. Robert R. Nathan Associates, Inc., *Institutional Implications of U.S. Deepwater Port Development For Crude Oil Imports*, (Washington, D.C.: June 1973), pp. 37-38.

3. Ibid., p. 39.

4. Ibid., pp. 6-7. The report also recommended that new legislation should serve additional purposes.

5. Statement of General Koisch to the Subcommittee on Investigations and Oversight, House Public Works Committee, June 17, 1971.

6. U.S. Senate, Special Joint Subcommittee on Deepwater Ports Legislation of the Committees on Commerce, Interior and Insular Affairs, and Public Works, "Deepwater Port Act of 1973," 93rd Congress, 1st session, July 24, 1973, pp. 292-94.

7. Ibid., p. 294.

8. Robert I. Price, "Anti-Pollution Measures—IMCO Subcommittee on Ship Design and Equipment," *Marine Technology*, Vol. 8, No. 1 (January 1971), pp. 1-7.

9. United States Coast Guard, *The Commandant's Bulletin*, 45-73 and 46-73 (November 1973).

10. Ibid.

11. Ibid.

Chapter 7
Labor Aspects of Deepwater Terminal Systems

1. Much of the material from this chapter is the result of numerous interviews with various government, labor, and industry personnel, as well as the combined experience and observations of the authors.

2. "More Oil Cargos for U.S. Tankers," *Business Week*, September 14, 1974, p. 33.

3. *Wall Street Journal*, June 5, 1973, p. 4; February 20, 1974, p. 8.

4. For example see: Samuel A. Lawrence, *United States Merchant Shipping Policies and Politics* (Washington: The Brookings Institute, 1966); Aaron Warner, et al., *Analysis of Labor-Management Relations Problems in the Offshore Operations of the East Coast Maritime Industry* (Washington: U.S. Maritime Administration, 1964); Joseph P. Goldberg, *Maritime Story, A Study of Labor Management* (Cambridge, Mass.: Harvard University Press, 1958).

5. William E. Simkin, "Effects of the Structure of Collective Bargaining in Selected Industries," *Proceedings of the 1970 Annual Spring Meeting*, Industrial Relations Research Association, pp. 514-15; Warner, pp. 54-55.

6. Warner, pp. 49-52.

7. Joseph P. Goldberg, "The Maritime Industry," *Proceedings of the 1970 Annual Spring Meeting*, Industrial Relations Research Association, pp. 506-7.

8. "To the Membership of the IOMM&P, on the Affiliation," *Master, Mate & Pilot* (Spring 1971), p. 32.

9. "Seafarers Union Said To Extract Money from Foreign Seamen for U.S. Politics," *Wall Street Journal*, June 30, 1970, p. 8; "Grand Jury Says Seafarers Made Illegal Donations," Ibid., July 1, 1970, p. 5.

10. "Senate Passes Energy Bill by A 42-to-28 Margin," *The Master, Mate & Pilot* (September 1974), p. 32.

11. "The Need for Deepwater Terminals," *Maritime, Voice of the AFL-CIO Maritime Trades Department* (Spring/Summer 1973), pp. 13-17.

12. "Gleason Calls on AFL-CIO to Halt MEBA-AMO Tactics," *The Master, Mate and Pilot* (March-April 1974).

13. "A Full Cargo of Optimism," *Newsweek*, August 12, 1974, p. 61.

14. Gary C. Raffaele and William R. Allen, "Manpower Problems Related to Large Supertankers and Offshore Terminals," *Decision Sciences Northeast Proceedings*, Vol. 3 (1974), pp. 40-43.

15. Louis K. Bragaw, William R. Allen, and Edward Roe, "The Impact of Environmental Regulation on the Marine Manager," *Proceedings of the American Institute for Decision Sciences*, Vol. 8 (1974), p. 31.

Chapter 8
Meeting the Challenge

1. The company that proposed to build one Gulf Coast deepwater port uses the Case III assumptions of x3.5 MBD without the adjustments for pipeline supplies. *Seadock*, p. 2.

2. Martin Fleck, "The Case for Deepwater Ports," *Exxon, U.S.A.*, Fourth Quarter, 1973, p. 14.

3. Robert Ciliano and Leonard Jacobson, "Tracing Secondary Environmental and Economic Impacts of Offshore Development Induced By Offshore Deepwater Superport Oil Terminal Facilities," *Decision Sciences Northeast Proceedings*, Vol. 3 (1974), pp. 34-36.

4. Arlen J. Large, "On Louisiana Coast, the Welcome Mat is Out for Giant Tanker Ships to Deliver Crude Oil," *Wall Street Journal*, October 30, 1973, p. 42.

5. Louis K. Bragaw, "Environmental Policy Formulation in Competitive Tanker Markets," *Decision Sciences Northeast Proceedings*, Vol. 3 (1974), pp. 31-33; "The Fear of Supertankers," (ed.), *Business Week*, July 27, 1974, p. 80.

6. U.S. Congress, House Committee on Public Works, Deepwater Ports, 93rd Congress, 1st Session, 1973, H. Rept. No. 93-66 to accompany H.R. 10701, p. 1.

Bibliography

Bibliography

Adelman, Morris A. "Is the Oil Shortage Real?" *Foreign Policy*, no. 9 (Winter 1972-73).

———. *The World Petroleum Market.* Baltimore, Md.: Johns Hopkins University Press, 1972.

Akins, James E. "The Oil Crisis: This Time the Wolf Is Here." *Foreign Affairs*, vol. 51, no. 3 (April 1973).

Baldwin, Malcom F. "Public Policy on Oil-An Econological Perspective." *Ecology Law Quarterly*, vol. 2 (Spring 1971).

Beman, Lewis. "Betting $20 Million in the Tanker Game." *Fortune* LXXXX, no. 20 (August 1974).

Benford, Harry. "Measures of Merit in Ship Design." *Marine Technology* 7, no. 4 (October 1970).

Berg, R.R., J.C. Calhoun, and R.L. Whiting. "Prognosis for Expanded U.S. Production of Crude Oil." *Science* 184, no. 4134 (April 19, 1974).

Boulding, Kenneth. "The Economics of Spaceship Earth," in *Environmental Quality in A Growing Economy*, ed. Henry Jarrett. Baltimore, Md.: Johns Hopkins University Press, 1972.

Bradley, Earl H., Jr., and John M. Armstrong. *A Description and Analysis of Coastal Zone and Shoreland Management Programs in the United States.* Ann Arbor, Mich.: University of Michigan, Sea Grant Program, Technical Report 20, 1972.

Bragaw, Louis K. "Environmental Policy Formulation in Competitive Tanker Markets." *Decision Sciences Northeast Proceedings*, vol. 3 (1974).

Bragaw, Louis K., William R. Allen, and Edward Roe. "The Impact of Environmental Regulation on the Marine Manager." *Proceedings of the American Institute for Decision Sciences*, vol. 8 (1974).

Bragaw, Louis K., and Leo Jordan. "U.S. Coast Guard Offshore Oil Pollution Control Systems." *Offshore Technology Conference Proceedings*, vol. I (May 1974).

Bragaw, Louis K., and Robert N. Stearns. "Environmental Costs and Optimal Tanker Size." XXI TIMS International Meeting, San Juan, P.R., October 18, 1974.

"Brotherhood of Marine Officers, Exxon US-Flag Fleet Tanker Officers Both Sign Affiliation Agreements with IOMM&P." *The Master, Mate and Pilot* (November 1974).

Ciliano, Robert, and Leonard Jacobson. "Tracing Secondary Environmental and Economic Impacts of Offshore Development Induced by Offshore Deepwater Superport Oil Terminal Facilities." *Decision Sciences Northeast Proceedings*, vol. 3 (1974).

"Corporate Advertising and the Environment." Report of Council on Economic Priorities, vol. 2, no. 3 (September 1971).

Degler, Stanley E. *Oil Pollution, Problems and Policies.* Washington, D.C.: Bureau of National Affairs, 1969.

Delaware River Port Authority, World Trade Division, Applied Research Bureau. *A Deepwater Port Analysis, Delaware River Bay.* Philadelphia, Pa.: Delaware River Port Authority, 1972.

Dillingham Corporation. *Systems Study of Oil Cleanup Procedures.* A Report to the Committee on Air and Water Conservation of the American Petroleum Institute, February 1970.

Energy Policy Project of the Ford Foundation. "Exploring Energy Choices: A Preliminary Report." Washington, D.C., 1974.

_____. *A Time to Choose.* Cambridge, Mass.: Ballinger Publishing Co., 1974.

"Fear of Supertankers, The." *Business Week* editorial, July 27, 1974.

Fearnley and Edgars. *Large Tankers.* Oslo, Norway: Fearnley and Edgars, 1974.

_____. *World Bulk Fleet.* Oslo, Norway: Fearnley and Edgars, 1974.

Fleck, Martin. "The Case for Deepwater Ports." *Exxon U.S.A.* Fourth Quarter, 1973.

Frankel, Bernard. "Offshore Tanker Terminals: Study in Depth." *U.S. Naval Institute Proceedings* (March 1973).

"Full Cargo of Optimism, A." *Newsweek*, August 12, 1974.

Garvey, Gerald. *Energy, Ecology, Economy.* New York, N.Y.: W.W. Norton & Co., 1972.

"Gleason Calls on the AFL-CIO to Halt MEBA-AMO Tactics." *The Master, Mate and Pilot* (March-April 1974).

Goldberg, Joseph P. "The Maritime Industry." *Proceedings of the 1970 Annual Spring Meeting.* Madison, Wisc.: Industrial Relations Research Association, 1970.

_____. *The Maritime Story.* Cambridge, Mass.: Harvard University Press, 1958.

Grad, Frank P. *Environmental Law.* New York, N.Y.: Mathew Bender & Company, 1971.

"Grand Jury Says Seafarers Made Illegal Donations." *The Wall Street Journal*, July 1, 1970.

Hayman, Christopher. "What to Do with All Those Tankers." *The New York Sunday Times*, April 14, 1974.

Hepple, Peter, ed. *Water Pollution by Oil.* London, England: Elsevier Publishing Co., 1971.

Hubbert, M. King. "The Energy Resources of the Earth." *Scientific American* 224, no. 3 (September 1971).

Hudson Engineers. *Preliminary Study of Six Mile Superport for Delaware Valley Council.* Philadelphia, Pa.: Hudson Engineers, September 1972.

Jantscher, Gerald R. *Bread upon the Waters: Federal Aid to the Maritime Industries.* Washington, D.C.: The Brookings Institution, 1974.

Jones, Lawrence G. "Just How Serious Was the Santa Barbara Oil Spill?" *Ocean Industry* (June 1969).

Kash, Don E., et al. *Energy Under the Oceans.* Norman, Okla.: University of Oklahoma Press, 1973.

Ketchem, Bostwick H. *The Water's Edge: Critical Problems of the Coastal Zone.* Cambridge, Mass.: The M.D.S. Press, 1972.

Kimon, Peter M., Ronald K. Kiss, and Joseph D. Porricelli. "Segregated Ballast VLCCs: An Economic and Pollution Abatement Analysis." Society of Naval Architects and Marine Engineers, Chesapeake Section, 1973.

Large, Alan J. "On Louisiana Coast, the Welcome Mat is Out for Giant Tanker Ships to Deliver Crude Oil." *The Wall Street Journal*, October 30, 1973.

Lawrence, Samuel A. *International Sea Transport: The Years Ahead.* Lexington, Mass.: Lexington Books, D.C. Heath and Company, 1972.

_____. *United States Merchant Shipping Policies and Politics.* Washington, D.C.: The Brookings Institution, 1966.

Like, Irving. "Multi Media Confrontation—The Environmentalist's Strategy for a No-Win Agency Proceeding." 1 *Ecology Law Quarterly* 495, 1971.

Maine State Planning Office Executive Department. *Marine Coastal Resources Renewal.* Augusta, Me., July 1972.

Marcus, Henry S. "The U.S. Superport Controversy." *Technology Review* 75, no. 5 (March/April 1973).

Moore, Steven, Robert L. Dwyer, and Arthur M. Katz. "A Preliminary Assessment of the Environmental Vulnerability of Machias Bay, Maine to Oil Supertankers." *Report No. MITSG 73-6.* Cambridge, Mass.: Massachusetts Institute of Technology, 1973.

"More Oil Cargoes for U.S. Tankers." *Business Week*, September 14, 1974.

Mostert, Noel. *Supership.* New York, N.Y.: Alfred A. Knopf, 1974.

Nathan Associates, Inc. "Guide to the National Petroleum Council Report on United States Energy Outlook." Washington, D.C., 1972.

_____. *Institutional Implications of the U.S. Deepwater Port Development for Crude Oil Imports.* Washington, D.C., June 1973.

_____. *U.S. Deepwater Port Study*, Vol. 1, Summary and Conclusions. Washington, D.C., August 1972.

National Petroleum Council, Committee on Emergency Preparedness. "Emergency Preparedness for Interruption of Petroleum Imports into the United States." Washington, D.C.: National Petroleum Council, 1974.

National Petroleum Council, Committee on Energy Outlook. "U.S. Energy Outlook." Washington, D.C.: National Petroleum Council, 1972.

"NEPA-Reform in Government Decision Making." Third Annual Report of Council on Environmental Quality, August 1972.

Odell, Peter. *Oil and World Power.* New York, N.Y.: Taplinger Publishing Co., 1970.

Oliver, E.F. "Gargantuan Tankers—Privileged or Burdened?" *U.S. Naval Institute Proceedings* (September 1970).

Porricelli, Joseph D., and Virgil F. Keith. "Tankers and the U.S. Energy Situation—An Economic and Environmental Analysis." ASME Publication

73-ICI-106. New York, N.Y.: The American Society of Mechanical Engineers, 1972.

"Port Growth Policies Abroad." *Water Spectrum* (Winter 1971-72).

Raffaele, Gary C., and William R. Allen. "Manpower Problems Related to Large Supertankers and Offshore Terminals." *Decision Sciences Northeast Proceedings*, vol. 3 (1974).

Raytheon Company. *Massport Marine Deepwater Terminal Study, Interim Report Phase IIA* (May 1974).

Regional and National Demands on the Maine Coastal Zone. Boston, Mass.: New England River Basins Commission, 1971.

Rosenthal, Albert J. "The Federal Power to Protect the Environment: Available Devices to Compel or Induce Desired Conduct." 45 *Southern California Law Review* 2, May 1972.

Rowe, James L., Jr. "Administration Weighs Tanker Industry Help." The Washington Sunday Post, May 11, 1975.

Sax, Joseph. "The Search for Environmental Quality: The Role of the Courts." *The Environmental Crisis.* New Haven, Conn.: Yale University, 1970.

"Seafarers Union Said to Extract Money from Foreign Seamen for U.S. Politics." *The Wall Street Journal*, June 30, 1970.

"Senate Passes Energy Bill by a 42-to-28 Margin." *The Master, Mate and Pilot* (September 1974).

Simkin, William E. "Effects of Structure of Collective Bargaining in Selected Industries." *Proceedings of the 1970 Annual Spring Meeting.* Madison, Wisc.: Industrial Relations Research Association, 1970.

Smith, F.A., et al. *Fourteen Selected Marine Resources Problems of Long Island, New York: Descriptive Evaluations.* Hartford, Conn.: The Travelers Research Corp., January 1970.

Smith, J.S. *Torrey Canyon Pollution and Marine Life.* London, England: Cambridge University Press, 1968.

Special Joint Subcommittee on Deepwater Ports Legislation of the Committees on Commerce, Interior and Insular Affairs, and Public Works. "Deepwater Port Act of 1973." 93rd Congress, 1st sess., July 24, 1973. Washington, D.C.: Government Printing Office, 1973.

Spencer, Janet. "Defending the Future: Sketches of Four Environmental Law Groups." *Juris Doctor* (February 1972).

"Statement of General Koisch to the Subcommittee on Investigations and Oversight." U.S. House of Representatives. Public Works Committee. June 17, 1971.

Stickells, Austin. *Federal Control of Business: Anti-Trust Laws.* Rochester, N.Y.: The Lawyers Co-operative Publishing Co., 1972.

Sturmey, S.G. *On the Price of Tramp Ship Freight Service.* Bergen, Norway: Institute for Shipping Research, 1965.

Testimoney of Max Blumer Before the Anti-Trust and Monopoly Subcommittee of the U.S. Senate. August 1970.

Todd, J.H., et al. "An Introduction to Environmental Enthology." Unpublished manuscript, WHDI-72-43. Woods Hole, Mass.: Woods Hole Oceanographic Institute, 1972.

"Tools for Coastal Zone Management." *Proceedings of the Conference of the Marine Technology Society.* Washington, D.C., 1972.

"To the Membership of the IOMM&P, on the Affiliation." *Master, Mate and Pilot* (Spring 1971).

U.S. Department of the Army, Corps of Engineers. *Deepwater Terminal in Delaware Bay.* Proceedings of Public Meeting, Philadelphia, Pa., 1970.

_____. "The Environmental and Ecological Aspects of Deepwater Ports." *U.S. Deepwater Port Study.* Vol. IV, August 1972.

U.S. Department of Commerce, Maritime Administration. *Deep Draft Vessel Port Capability on the U.S. North Atlantic Coast.* Washington, D.C.: Government Printing Office, 1970.

U.S. Department of the Interior. "An Analysis of the Economic and Security Aspects of the Trans-Alaska Pipeline." Vols. I-II. Washington, D.C.: Government Printing Office, 1971.

_____. *Minerals, Facts and Problems, 1970.* Bureau of Mines Bulletin 650. Washington, D.C.: Government Printing Office, 1970.

U.S. Department of Transportation. *Refinery-Deepwater Port Location Study.* Washington, D.C.: Department of Transportation, 1974.

U.S. Department of Transportation, Coast Guard. *Polluting Spills in the United States Waters, 1970.* Washington, D.C.: U.S. Coast Guard, September 1971.

U.S. House of Representatives. Committee on Public Works. "Deepwater Ports." House Report No. 93-66 to accompany H.R. 10701. 93rd Congress, 1st sess. Washington, D.C.: Government Printing Office, 1973.

U.S. Senate. "Deepwater Port Act of 1974." 93rd Congress, 2nd sess. Washington, D.C.: Government Printing Office, 1974.

Warner, Aaron, Alfred Eickner, and James J. Healy. *Analysis of Labor Management Relations Problems in the Off-Shore Operations of the East Coast Maritime Industry.* Washington, D.C.: U.S. Maritime Administration, 1974.

Wiggins, B. "Applying the Fairness Doctrine to Environmental Issues." *Natural Resources Journal*, vol. 12 (January 1, 1972).

Zannetos, Zenon S. *The Theory of Oil Tankship Rates.* Cambridge, Mass.: The M.I.T. Press, 1966.

Index

Index

Adelman, Morris, 17, 19, 20
American Coal Shipping Company, 116
American Federation of Labor-Congress of Industrial Organizations (AFL-CIO), 114, 116, 117, 118
American Maritime Association (AMA), 117
American Merchant Marine Institute (AMMI), 117
American Radio Association (ARA), 115
American Tobacco Company, 81
American Trading and Production Company, 116
Analysis of the Economic and Security Aspects of the Trans-Alaska Pipeline, 6
Anglo-Norness, Inc., 61
Arthur D. Little, Inc., 86, 93
Associated Maritime Officers (AMO), 117, 118, 120-122
Associated Press, 81
Associated Press v. United States, 81

Ballast designs, 60
Bender, C.R., 106
Blumer, Max, 70
Boulding, Kenneth, 11
Britain, 22
Brotherhood Marine Officers (BMO), 116, 117, 118, 120-122
Bureau of Mines, 4

Catenary anchor leg systems (CALM), 46-49
Clayton Act, 81, 82
Clean Air amendments, 78
Coastal Conveyance of Petroleum Act, 77
Coastal Wetland Regulation Act, 77
Coastal zone, defined, 76
Coastal Zone Management Act, 76, 77, 78, 83, 90
Commission on Marine Science, 85
Continental Shelf Convention, 96
Council of Economic Advisors, 41
Council on Environmental Quality, 84

Crapp, 69
Cross, James V., 101
Curran, Joseph, 116, 118

Deepwater Port Act, 99, 101-105, 113, 135
Deepwater terminals, 1-2, 12; alternatives to, 133; costs, 49-52; economic analysis of impact of, 105; economic aspects of, 41-60, 132-133; and environmental concerns, 61-62, 74, 132-133; legal issues involving, 74-82; locations of, 27-39, 76, 130-131; and manpower issues, 124-127; as an operating system, 24-25; and political issues, 82-86; and socioeconomic issues, 86-87; types of, 43-49, 100; unions and, 109-124; within territorial seas of U.S., 96
Delaware Coastal Zone Act, 80
Department of Transportation, 89, 101, 104, 105
Douglas, William O., 79
Dyckerhoff & Widmann, 43

Ecology, 132-133; offshore terminals and environmental considerations, 61-64; and tankers, 73-74, 131-132; United States policy on environmental issues, 84. *See also* Pollution
Economics of Deepwater Terminals, The, 53
Economy, 132-133; of deepwater terminals, 41-60
Energy, 1; growth of demand, 3-4, 6; potential sources of, 2; and self-sufficiency, 129-130; supply and demand, 7-12
Energy Policy Project, 13; study of, 9-12
Energy Transportation Security Act, 113, 114, 119, 120, 121, 123-124
Exploring Energy Choices, 9
Extension of Admirality Jurisdiction Act, 77-78
Exxon, 114, 122, 123

Fearnley and Egers Chartering Company, Ltd., 23

159

About the Authors

Louis K. Bragaw is Professor of Management and Head of the Department of Economics and Management at the United States Coast Guard Academy, New London, Connecticut. His teaching, research, and consulting interests focus on government and business policy and planning and the management of the policy-making and resource allocation processes. During 1975-76 he is an Eli Lilly faculty fellow in the Public Policy and Decision Making Program at Yale University, and coeditor of the Eastern Academy of Management *Proceedings*. He is the author of over twenty-five articles, papers and case studies. Dr. Bragaw received the D.B.A. from the George Washington University, the M.S. and the Naval Engineer degree from the Massachusetts Institute of Technology, the M.B.A. from Northeastern University, and the B.S. from the United States Coast Guard Academy.

Henry S. Marcus is Assistant Professor of Marine Systems in the Department of Ocean Engineering and Executive Officer of the Commodity Transportation and Economic Development Laboratory at the Massachusetts Institute of Technology. His experience includes consulting on a wide variety of problems in the area of transportation and logistics. The author or coauthor of two other books in the maritime field, his present activities include research on the federal role in port planning. Dr. Marcus received the D.B.A. from Harvard Graduate School of Business Administration, two M.S. degrees from the Massachusetts Institute of Technology, and the B.S. from the Webb Institute of Naval Architecture.

Gary C. Raffaele is Associate Professor of Management in the College of Business Administration at the University of Texas at San Antonio. His research interests focus on industrial relations and manpower development. A licensed deck officer, he sailed on various tankers and other ships for five years before completing his graduate studies. Dr. Raffaele received the D.B.A. from Harvard Graduate School of Business Administration, the M.B.A. from the University of Texas, and the B.S. from the State University of New York, Maritime College.

James R. Townley is a member of the Planning Staff of the Office of Merchant Marine Safety, United States Coast Guard Headquarters, Washington, D.C. He served aboard the Coast Guard Cutters *Campbell* and *Morgenthau* and has been a marine inspector and resident construction inspector at a major shipyard during the construction of several VLCCs. Mr. Townley received the M.S. in shipping and shipbuilding management from the Massachusetts Institute of Technology, the M.B.A. from New York University, and the B.S. from the United States Coast Guard Academy.